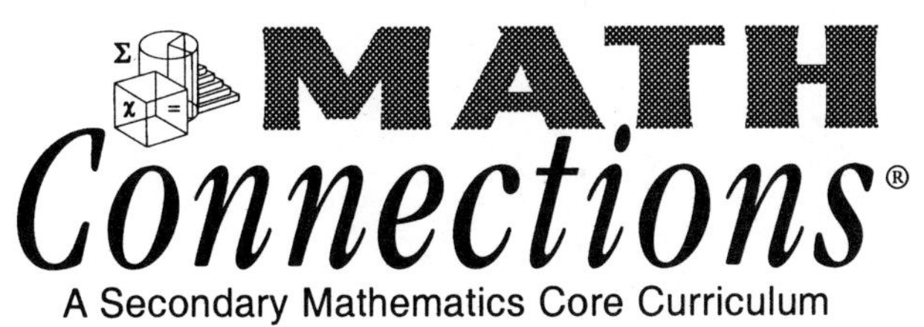

Student Assessments with Solution Keys and Scoring Guides

The assessments for this text, **MATH** *Connections* Year I, reflect the intent of the NCTM Standards on Assessment in several ways.

- The assessments allow students to use a variety of formats to demonstrate their knowledge of mathematics concepts and skills. Students are consistently required to analyze problems, determine solutions and write justifications for their solutions. Some quizzes have performance assessment items that may be assigned to a group or to an individual.
- The assessments are tied to specific mathematics objectives. Each assessment item is directly linked to the Learning Outcomes for each section of the chapter. The assessment items on the chapter test may not match every learning outcome for a chapter.
- The assesssments focus on communication, reasoning, problem solving and connections as well as skill development. They reflect the concepts, content and processes of mathematics.
- The Solution Key and Scoring Guide for each quiz and chapter test contains solutions to all questions as well as *suggestions* for scoring based on a 100 point scale. A rubric has been developed for open-ended questions, which details the elements of a rich response.

Two assessment forms, Form A and Form B, were developed for each quiz and chapter test. The two forms are designed to be of comparable difficulty and provide you with an alternative assessment for each quiz and chapter test.

— Donald Hastings
Assessment Specialist
Stratford Public Schools (retired)

Published 2000 by IT'S ABOUT TIME, Inc., Armonk, New York.
Previously published for field-testing © 1992, 1996.

Assessments Form A

Assessments Form A & B — Are designed to allow students to use a variety of formats to demonstrate their knowledge of mathematics concepts and skills. Assessments are tied to specific mathematics objectives and are linked to the Learning Outcomes for each section of the chapter. Two Assessment forms, Form A and Form B, were developed for each quiz and chapter test. The Solution Key and Scoring Guide for each quiz and chapter test contains solutions to all questions as well as suggestions for scoring based on a 100 point scale.

Legend:

The header to each Assessment is explained as follows:

Upper left corner: **MATH** *Connections* I refers to Year 1.

Upper right corner: Quiz 1.1–1.2 (A) refers to a quiz for Chapter 1, Sections 1.1–1.2, Form A. The reference sk: Quiz: 1.1–1.2 (A) refers to the Solution Key and Scoring Guide for the Quiz for Chapter 1, Sections 1.1–1.2, Form A.

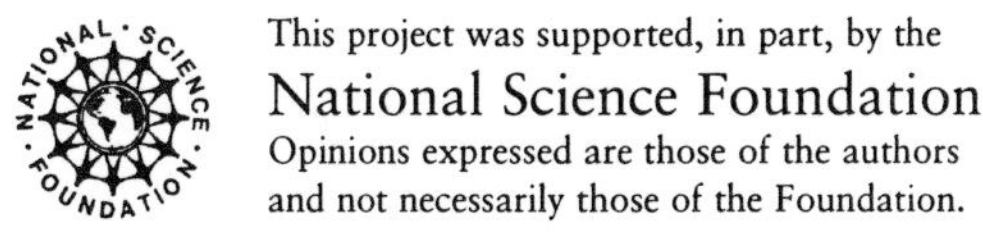

This project was supported, in part, by the
National Science Foundation
Opinions expressed are those of the authors and not necessarily those of the Foundation.

ISBN 1-891629-33-6
ISBN 1-58591-030-9
Published by IT'S ABOUT TIME, Inc.

MATH *Connections* **I**
TURNING FACTS INTO IDEAS QUIZ
SECTIONS 1.1 - 1.2 (A)

Name ______________________________ Date____________________

1. The following data deals with sneaker prices of popular brands of sneakers sold at Karl's Discount Store:

SNEAKER PRICES	
Air Jordan by Nike	$125
Air Macs by Nike	$110
Air 180 by Nike	$125
Air Tech Challenge by Nike	$85
Asics Running Shoe	$85
Blacktop by Reebok	$70
Bo Jackson by Nike	$110
Converse All-Star	$30
Ex-O-Fit by Nike	$60
Flight by Nike	$80
Force by Nike	$75
Pump by Reebok	$125
Puma	$80

a. Compute the mean for the sneaker prices to the nearest dollar. _______

b. Construct a dotplot for the sneaker prices.

____

c. What is the frequency of sneakers that cost $85.00? _____

d. What is the mode for the prices of the sneakers listed in the table? ___________

e. Explain why knowing the mean for sneaker prices could influence your decision to buy a particular brand of sneaker.

____

____

____

f. When buying a pair of sneakers what features do you consider the most important?

____

____

____

____

2. Thomas Upright is president of the Merryweather Paper Company. He asks you to develop a bar graph using the table for the employee absences for the past five weeks. The table is displayed below:

MERRYWEATHER PAPER COMPANY
ABSENTEEISM 720 EMPLOYEES (TOTAL)

WEEK	M	T	W	TH	F
1	10	5	4	3	8
2	8	6	4	3	9
3	7	6	5	2	11
4	9	5	6	3	12
5	5	4	3	1	10

a. Use this information to construct a bar graph. Let the bars show the total number of absences for each day during the five weeks.

Use either the table or your bar graph to answer the following questions:

b. What are the total number of absences during the five weeks? _______

c. How many more people were absent on Fridays than Thursdays in the five weeks? _____

d. Why do you think there is such a difference in absences on Friday compared to Thursday?

____

____

____

____

e. If you were the president of the company, what policy changes would you make to change the absentee pattern?

____

____

____

____

3. Which was more helpful in answering these questions, the table or bar graph? Explain your answer.

____

____

____

____

4. a. Compute the mean for the weekly absences by dividing the total number of absences by 5. ________

   b. Does this mean give you any helpful information that you could use to answer the questions posed by the president? Explain your answer.

____

____

5. a. Describe three different methods for collecting data.

____

____

____

____

____

b. Describe any difficulties one may have in collecting data.

____

____

____

____

SOLUTION KEY AND SCORING GUIDE

MATH *Connections* **I**
TURNING FACTS INTO IDEAS QUIZ
SECTIONS 1.1 - 1.2 (A)

1. The following data deals with sneaker prices of popular brands of sneakers sold at Karl's Discount Store:

SNEAKER PRICES	
Air Jordan by Nike	$125
Air Macs by Nike	$110
Air 180 by Nike	$125
Air Tech Challenge by Nike	$85
Asics Running Shoe	$85
Blacktop by Reebok	$70
Bo Jackson by Nike	$110
Converse All-Star	$30
Ex-O-Fit by Nike	$60
Flight by Nike	$80
Force by Nike	$75
Pump by Reebok	$125
Puma	$80

a. Compute the mean for the sneaker prices to the nearest dollar. *$89.00* ***(6)***

b. Construct a dotplot for the sneaker prices. ***(9)***

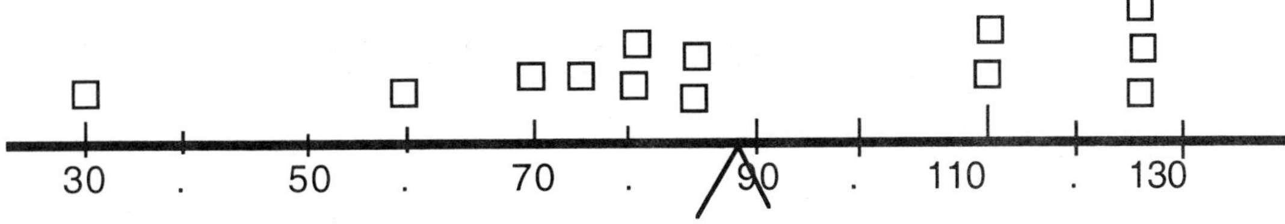

c. What is the frequency of sneakers that cost $85.00? *2* ***(6)***

d. What is the mode for the prices of the sneakers listed in the table? *$125* ***(6)***

e. Would knowing the mean of sneaker prices influence your decision to buy a particular brand of sneaker? Explain your answer ***(6)***

A rich answer should contain the following elements:

- *The mean is the average price of the sneakers.*
- *Outliers affect the mean.*
- *A discussion is made about the number of sneaker prices above and below the mean.*
- *A discussion is made about the spread of the data.*

A score of 6 would be achieved if the student included at least three items with some elaboration. Scores of 4 or 5 would be achieved if the student named three items with a minimum of elaboration. A score of 3 would be achieved if the student named less than three items with some elaboration. Scores of 2 or 1 would be achieved if the student named 1 or 2 items with no elaboration. A score of 0 is given for no response.

f. When buying a pair of sneakers what features do you consider the most important? ***(6)***

Scoring an open-ended, opinion type question will depend heavily on the level of the student and the expectations of the teacher. Since many students will have limited experience writing responses to open-ended questions, it is important to share your expectations with students. A rich answer would include complete sentences, correct grammar and spelling, thoughtful comments and comparative information. At the beginning of the year, a teacher may use this type of question for extra credit. By the end of the year, it is expected that all students can handle questions of this type.

2. Thomas Upright is president of the Merryweather Paper Company. He asks you to develop a bar graph using the table for the employee absences for the past five weeks. The table is displayed below:

MERRYWEATHER PAPER COMPANY
ABSENTEEISM (720 EMPLOYEES TOTAL)

WEEK	M	T	W	TH	F
1	10	5	4	3	8
2	8	6	4	3	9
3	7	6	5	2	11
4	9	5	6	3	12
5	5	4	3	1	10

a. Use this information to construct a bar graph. Let the bars show the total number of absences for each day during the five weeks. ***(9)***

EMPLOYEE ABSENTEEISM AT THE MERRYWEATHER PAPER COMPANY
WILLIMANTIC DIVISION

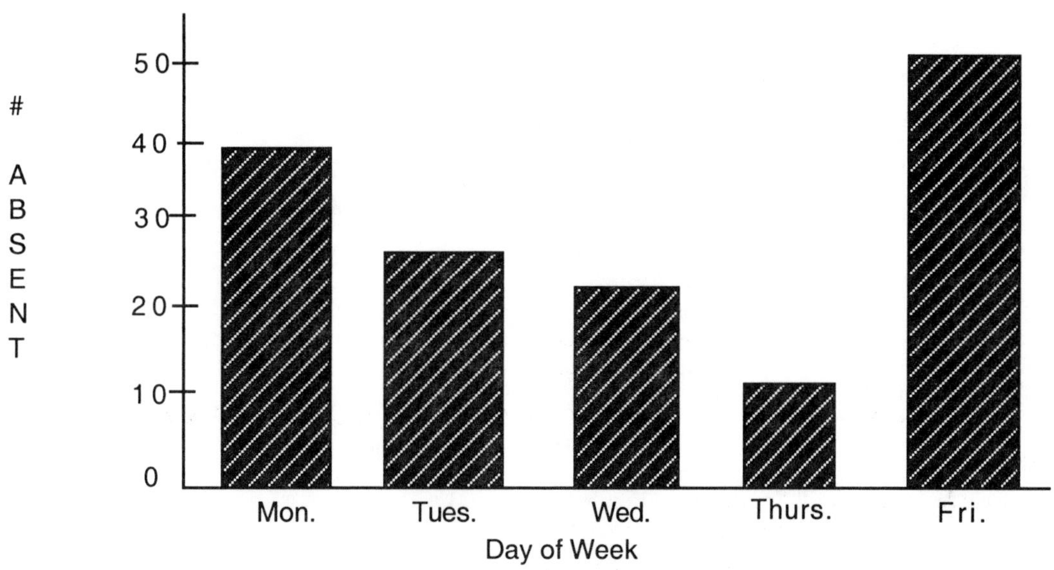

Use either the table or your bar graph to answer the following questions.

b. What are the total number of absences in the five weeks? *149* ***(6)***

c. How many more people were absent on Fridays than Thursdays during the five weeks? *38* ***(6)***

d. Why do you think there is such a difference in absences on Friday compared to Thursday? ***(6)***

A rich answer might include reasons such as:

- *Payday is on Thursday and ......(an elaboration of ideas)*
- *Workers like to have long weekends because .....(an elaboration of ideas)*
- *The work is difficult or boring so that one day a week absence is needed for mental health*

e. If you were the president of the company, what policy changes would you make to change the absentee pattern? ***(6)***

A rich answer might include two or more of the following:

- *The president could make Friday payday.*
- *The president could give incentives for workers with perfect attendance.*

- *The president could form a workers council to listen to ways to make the work more interesting and to improve working conditions.*

3. Which was more helpful in answering these questions, the table or bar graph? Explain your answer. ***(6)***

   The response should deal with the following considerations:
   - *The table is more convenient when you need specific information, you need to count or when you need to do computations.*
   - *The bar graph is more convenient when you are looking for trends.*

4. a. Compute the mean for the weekly absences by dividing the total number of absences by 5. <u>*29.8 or 30*</u> ***(5)***

   b. Does this mean give you any helpful information that you could use to answer the questions posed by the president? Explain your answer. ***(5)***

   No. This mean gives no indication on which days of the week the absences occur.

5. a. Describe three different methods for collecting data. ***(6)***

   A rich answer would include some of the following:
   - *Surveys*
   - *Newspapers, magazines, almanacs, television, radio*
   - *Experiments, observations*
   - *Internet*
   - *Information about people from school, medical records, public offices*

   b. Describe any difficulties one may have in collecting data. ***(6)***

   A rich answer could include some of the following:
   - *People who respond to questions may not tell the truth.*
   - *Information in media may be biased.*
   - *Experiments may be improperly developed.*
   - *Public Information may be confidential.*

MATH *Connections* **I**
TURNING FACTS INTO IDEAS QUIZ
SECTION 1.3 (A)

Name ________________________________ Date ____________________

1. a. List all the numbers that are displayed in the stem-and-leaf plot below:

Stem	Leaf
0	5
1	2 9 3 6 3
2	4 0 5
3	9 0 2 9 1
4	1 3

3 | 9 means 39

b. Find the mean for the data to the nearest whole number. ________

2. Sabia Ali is a worker for the United Nations and wishes to make a report on the percentage of family income spent on food that is consumed in the home. She has the following data available to her:

PERCENT OF INCOME SPENT ON FOOD CONSUMED AT HOME

COUNTRY	PERCENT	COUNTRY	PERCENT
UNITED STATES	10	MALAYSIA	25
AUSTRIA	16	NETHERLANDS	14
BELGIUM	16	NORWAY	19
CANADA	11	PHILIPPINES	53
COLOMBIA	28	SINGAPORE	18
DENMARK	15	SOUTH AFRICA	27
ECUADOR	32	SPAIN	22
FRANCE	16	SRI LANKA	50
WEST GERMANY	20	SUDAN	63
GREECE	31	SWEDEN	16
HONDURAS	41	SWITZERLAND	19
INDIA	51	THAILAND	27
ITALY	19	UNITED KINGDOM	12
LUXEMBOURG	13	VENEZUELA	36
		ZIMBABWE	13

a. Sabia asks you to construct a stem-and-leaf plot of the data. Do that below.

b. In order to analyze the data further, it is decided to enter the data into a calculator by using the STAT menu and the EDIT option. The data is entered in L_1.

c. From the STAT menu and CALC option, the 1-VAR STATS choice is made. From the list you select the following information:

$\bar{x}$ ________

Σx _______

n ________

d. Sabia asks you to identify in words what the three symbols mean. You tell her:

$\bar{x}$ ____

Σx____

n ____

e. Sabia asks you to explain the relationship among $\bar{x}$, Σx and n. You tell her:

____

____

____

f. Sabia asks you to tell her how many countries are above the mean. You tell her: ________

g. Sabia asks you to use a calculator to draw a histogram for this data. To do this you need to set the WINDOW. Your choices are:

Xmin ______ Ymin _____

Xmax_____ Ymax _____

Xscl _______ Yscl _____

Then you select the STAT PLOT MENU and choose PLOT 1, turning it on, selecting histogram, L_1 and frequency of 1. Sketch the histogram for Sabia below:

h. Sabia is not sure which display to use in her report.

- She asks you to give one reason why a stem-and-leaf plot is good for displaying this data. You tell her:

  ____

  ____

- She asks you to give one reason why a histogram is good for displaying this data. Your answer is:

  ____

  ____

3. Construct a set of data with ten or less items for which there will be fewer data items above the mean for the data than below the mean.

SOLUTION KEY and SCORING GUIDE

MATH *Connections* **I**
TURNING FACTS INTO IDEAS QUIZ
SECTION 1.3 (A)

1. a. List all the numbers that are displayed in the stem-and-leaf plot below:***(10)***

0	5
1	2 9 3 6 3
2	4 0 5
3	9 0 2 9 1
4	1 3

3 | 9 means 39

5, 12, 19, 13, 16, 13, 24, 20, 25, 39, 30, 32, 39, 31, 41, 43,

Find the mean for the data to the nearest whole number. ___*25*___***(5)***

2. Sabia Ali is a worker for the United Nations and wishes to make a report on the percentage of income spent on food that is consumed in the home. She has the following data available to her:

PERCENT OF INCOME SPENT ON FOOD CONSUMED AT HOME

COUNTRY	PERCENT	COUNTRY	PERCENT
UNITED STATES	10	MALAYSIA	25
AUSTRIA	16	NETHERLANDS	14
BELGIUM	16	NORWAY	19
CANADA	11	PHILIPPINES	53
COLOMBIA	28	SINGAPORE	18
DENMARK	15	SOUTH AFRICA	27
ECUADOR	32	SPAIN	22
FRANCE	16	SRI LANKA	50
WEST GERMANY	20	SUDAN	63
GREECE	31	SWEDEN	16
HONDURAS	41	SWITZERLAND	19
INDIA	51	THAILAND	27
ITALY	19	UNITED KINGDOM	12
LUXEMBOURG	13	VENEZUELA	36
		ZIMBABWE	13

a. Sabia asks you to construct a stem-and-leaf plot of the data. Do that below. ***(12)***

1	0 6 6 1 5 6 9 3 4 9 8 6 9 2 3
2	8 0 5 7 2 7
3	2 1 6
4	1
5	1 3 0
6	3

5|1 means 51

b. In order to analyze the data further it is decided to enter the data into a calculator by using the STAT menu and the EDIT option. The data is entered in L_1.

c. From the STAT menu and CALC option, the 1-VAR STATS choice is made. From the list you select the following information: ***(6)***

$\bar{x}$ *25.27*

Σx *733*

n *29*

d. Sabia asks you to identify in words what the three symbols mean. You tell her: ***(6)***

$\bar{x}$ *The mean*

Σx *The sum of the data values or the sum of the percentages*

n *The number of countries*

e. Sabia asks you to explain the relationship among x, Σx and n. You tell her: ***(10)***

$\bar{x}$ is the mean and is computed by dividing Σx by n.

f. Sabia asks you to tell her how many countries are above the mean. You tell her: *11* ***(5)***

g. Sabia asks you to use a calculator to draw a histogram for this data. To do this you need to set the WINDOW. Your choices are: ***(12)***

Xmin ______ Ymin _____

Xmax______ Ymax _____

Xscl _______ Yscl _____

The choices will differ. To receive full credit the window must accommodate the histogram.

Then you select the STAT PLOT MENU and choose PLOT 1, turning it on, selecting histogram, L_1 and frequency of 1. You sketch the histogram for Sabia below: ***(10)***

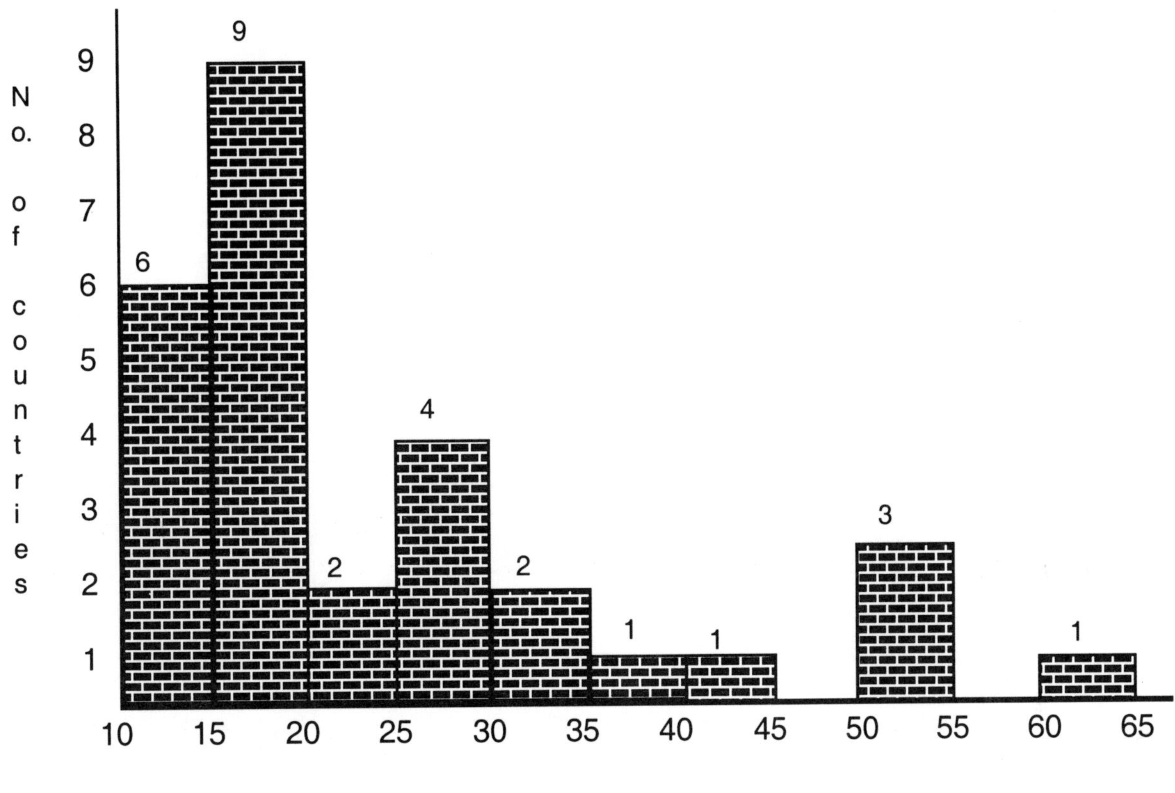

h. Sabia is not sure which display to use in her report. She asks you to give one reason why a stem-and-leaf plot is good for displaying this data and one reason why a histogram is good for displaying this data. She will make the decision then. Your answer is: ***(12)***

The response should include:
A stem-and-leaf is good because (1) we can read the actual values from it and (2) if we turn it we also have a histogram.
The histogram is good because: (1) we can read the frequencies for a given interval very quickly and (2) most people know how to read a histogram.

3. Construct a set of data with ten or less items for which there will be fewer data items above the mean for the data than below the mean. ***(12)***

An example: 12, 13, 14, 15, 50
The mean is 20.8. There is only one data item above the mean and 4 below.

MATH *Connections* **I**
TURNING FACTS INTO IDEAS QUIZ
SECTIONS 1.4 - 1.5 (A)

Name ______________________________ Date ______________

1. Tom Roberts monitored the number of adult salmon that returned through the Rainbow Fishway each year from 1978 to 1992. The boxplot for the data that Tom collected is pictured below:

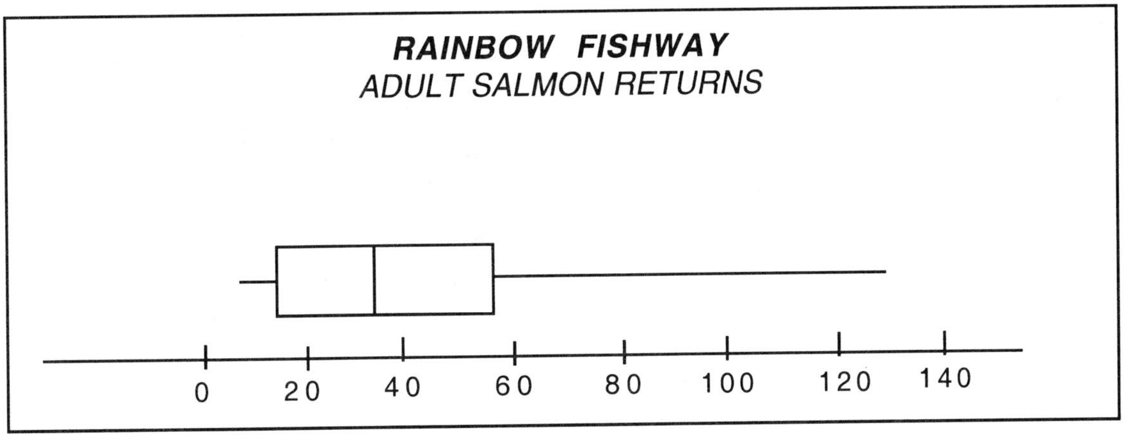

RAINBOW FISHWAY
ADULT SALMON RETURNS

0 20 40 60 80 100 120 140

a. Why is one whisker on this boxplot longer than the other?

____

____

____

b. What is the median number of adult salmon that returned to the Rainbow Fishway?

____

c. Why is the median not in the middle of the box?

____

____

____

d. When Tom records the number of salmon that returned in 1993, what number of fish is he likely to record if 1993 turns out to be an "average" year?

____

2. Linda is training for a triathlon and needs to monitor her carbohydrate intake in order to increase her stamina. She looks in a recent *The World Almanac and Book of Facts* for the carbohydrate content in various fruits to assist in planning her diet. The table below is the result of her research:

CARBOHYDRATES IN FRUIT (by grams)

Apple, raw, 2.75 in. diam.	20	Cantaloupe, 5 in. diam.	20
Applejuice	30	Orange, 2.625 in. diam.	16
Applesauce, canned, sweetened	61	Orange juice, frozen, diluted	29
Apricots, raw	14	Peach, raw, 2.5 in. diam.	10
Banana, raw	26	Peaches, canned in syrup	51
Cherries, sweet, raw	12	Pear, raw, 2.5 in. diam	25
Fruit cocktail, canned, heavy syrup	50	Pineapple, heavy syrup	49
Grapefruit, raw, med, white	12	Raisins, seedless	112
Grapes, Thompson seedless	9	Strawberries, whole	31
Lemonade, frozen, diluted	28	Watermelon, 4 by 8 in. wedge	30

a. Using a calculator, select the STAT menu and the EDIT option. Enter the data in L_1.

b. In the STAT menu choose CALC and the 1-VARIABLE STATS option to find the mean for this data. __________

c. If Linda wants to eat fruits that have carbohydrate quantities above the mean, which fruits should she choose?

____

____

d. Do you think that would be a wise choice? _____ Explain your answer.

____

____

____

____

e. Put the data in ascending order on the calculator. What is $L_1(1)$? ________
What is $L_1(10)$? ___

f. Use the ordered data to make a stem-and-leaf plot.

g. Use the stem-and-leaf plot to find the median and the mode(s).

median ____ mode(s) ____________

h. Use the stem and leaf plot to determine the five-number summary:

Min x ________ Quartile 1 _________

Max x ________

Median _______ Quartile 3 ___________

i. The number of carbohydrates in raisins, 112, appears to be an outlier. Explain how the 112 affects the median and mean.

____

____

____

____

j. On a calculator use the STAT PLOT function to display the boxplot for this data set. Copy the boxplot onto your test paper.

Use the TRACE function to check the five-number summary in problem (f). If any answer on the calculator is different, write it in parentheses next to your answer in (h).

3. The boxplots below summarize the Nielsen ratings for one summer week in 1989.

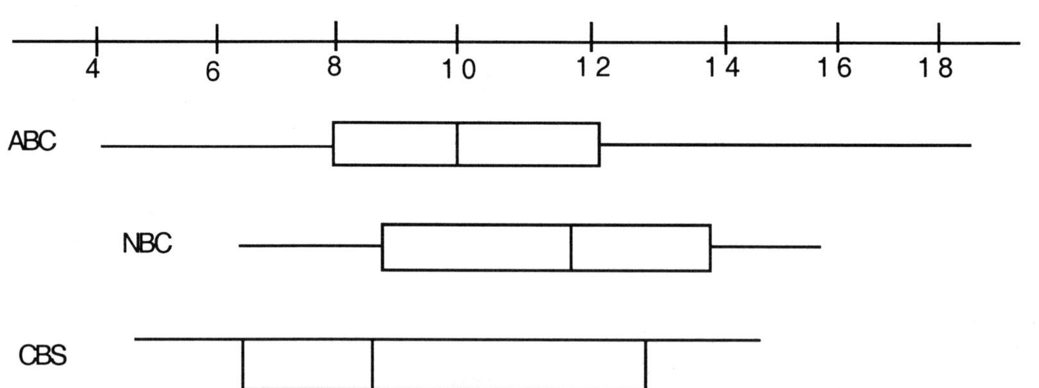

a. If winning the ratings for the week meant you had to have the highest rated show, which network would have won the ratings for that week? ________

b. What network had the lowest rated show? ________

c. What network had the highest median? __________

d. If winning the ratings for the week meant you had to have the highest third quartile, which network would have won the ratings for that week? ______

e. Which network do you think won the ratings battle for the week? ______ Explain.

____

____

____

____

f. Which network had the largest spread in its ratings? ________ Explain how you made your choice.

____

____

____

____

g. Officials at CBS were very unhappy when the ratings were announced. Explain why.

____

____

____

____

h. CBS decided to cancel the 4 shows with the lowest ratings. After cancelling the 4 shows their 5-number summary was: min: 6.9; first quartile: 9.1; median: 10.85; third quartile: 13.6; max: 14.6. Did cancelling the 4 shows with the lowest ratings help? ____ Explain.

____

____

____

4. a. Explain what the coach means by his statement?

____

____

____

____

____

____

b. Is the mean or median the best measure of central tendency for this example? Justify your answer.

____

____

____

____

____

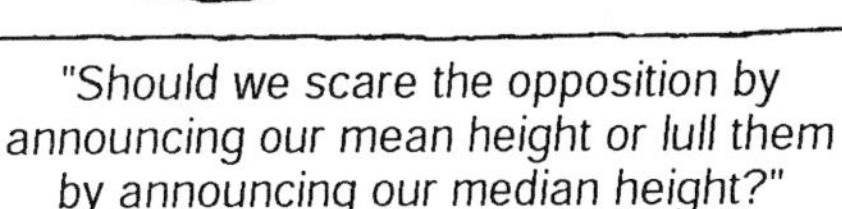

"Should we scare the opposition by announcing our mean height or lull them by announcing our median height?"

SOLUTION KEY AND SCORING GUIDE

MATH *Connections* **I**
TURNING FACTS INTO IDEAS QUIZ
SECTIONS 1.4 - 1.5 (A)

1. Tom Roberts monitored the number of adult salmon that returned through the Rainbow Fishway each year from 1978 to 1992. The boxplot for the data that Tom collected is pictured below:

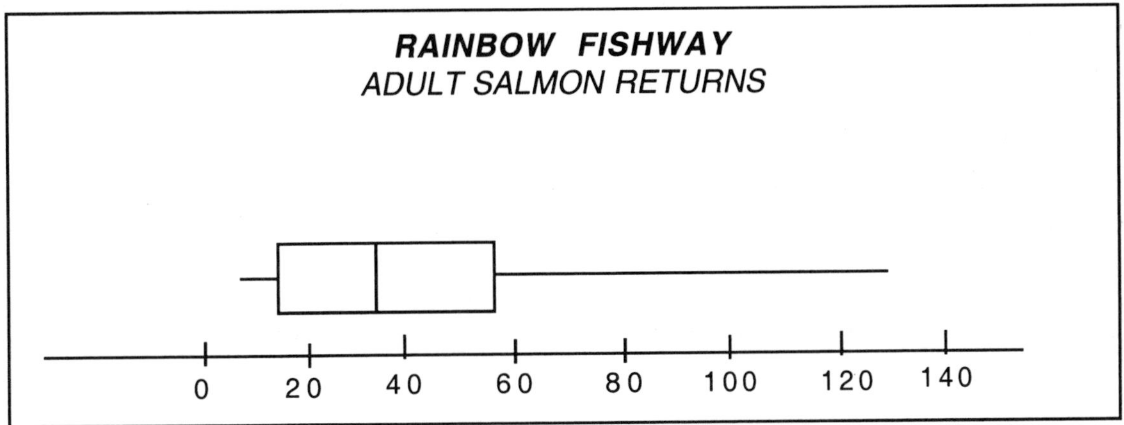

a. Why is one whisker on this boxplot longer than the other? ***(3)***

There was more spread in the top 25% of the data than in the lowest 25% of the data.

b. What is the median number of adult salmon that returned to the Rainbow Fishway?
about 35 ***(3)***

c. Why is the median not in the middle of the box? ***(3)***

There is more spread from the median to Q3 than from Q1 to the median.

d. When Tom records the number of salmon that returned in 1993, what number of fish is he likely to record if 1993 turns out to be an "average" year? ***(3)***

Any number from 14 to 55, since these data are the most representative. That is the middle 50%.

2. Linda is training for a triathlon and needs to monitor her carbohydrate intake in order to increase her stamina. She looks in a recent *The World Almanac and Book of Facts* for the carbohydrate content in various fruits to assist in planning her diet. The table below is the result of her research:

CARBOHYDRATES IN FRUIT (by grams)			
Apple, raw, 2.75 in. diam.	20	Cantaloupe, 5 in. diam.	20
Applejuice	30	Orange, 2.625 in. diam.	16
Applesauce, canned, sweetened	61	Orange juice, frozen, diluted	29
Apricots, raw	14	Peach, raw, 2.5 in. diam.	10
Banana, raw	26	Peaches, canned in syrup	51
Cherries, sweet, raw	12	Pear, raw, 2.5 in. diam	25
Fruit cocktail, canned, heavy syrup	50	Pineapple, heavy syrup	49
Grapefruit, raw, med, white	12	Raisins, seedless	112
Grapes, Thompson seedless	9	Strawberries, whole	31
Lemonade, frozen, diluted	28	Watermelon, 4 by 8 in. wedge	30

a. Using a calculator, select the STAT menu and the EDIT option. Enter the data in L_1.

b. In the STAT menu choose CALC and the 1-VARIABLE STATS option to find the mean for this data. ___*31.75*___ ***(2)***

c. If Linda wants to eat fruits that have carbohydrate quantities above the mean, which fruits should she choose? *Pineapple in heavy syrup, fruit cocktail in heavy syrup, peaches in syrup, canned applesauce, raisins.* ***(5)***

d. Do you think that would be a wise choice? Explain your answer. ***(6)***
These answers can vary greatly. A "no" answer may follow with reasoning that the above choices have too much sugar. A "yes" answer may follow with reasoning that Linda is receiving the wanted carbohydrates and sugar for quick energy. This is a problem that requires the student to take a stand and defend an answer.

e. Put the data in ascending order on the calculator. What is $L_1(1)$? ___*9*___ What is $L_1(10)$? ___*26*___ ***(2)***

f. Use the ordered data to make a stem-and-leaf plot.

0	9
1	0 2 2 4 6
2	0 0 5 6 8 9
3	0 0 1
4	9
5	0 1
6	1
11	2

1|0 means 10 gm of carbohydrates

(4)

g. Use the stem-and-leaf plot to find the median and the mode(s).

median *27* mode(s) *12; 20; 30* ***(3)***

h. Use the stem-and-leaf plot to determine the five-number summary: ***(5)***

Min x *9*

Max x *112*

Median *27*

Quartile 1 *15*

Quartile 3 *40*

i. The number of carbohydrates in raisins, 112, appears to be an outlier. Explain how the 112 affects the median and mean.

The 112 has the affect of making the mean higher. It has no effect on the median.

j. On a calculator use the STAT PLOT function to display the boxplot for this data set. Copy the boxplot onto your test paper. ***(4)***

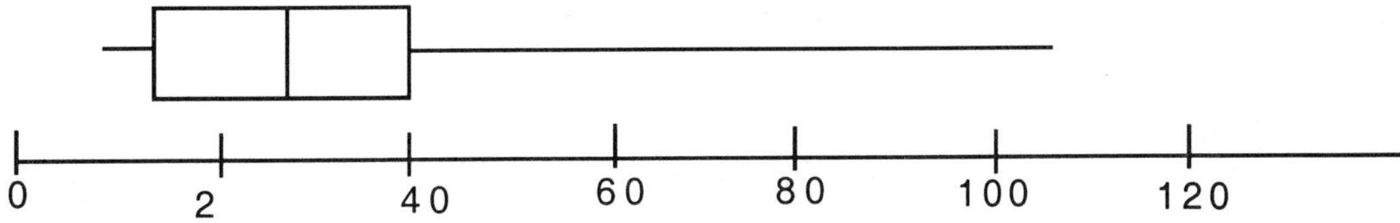

Use TRACE function to check the five-number summary in problem (f). If any answer on the calculator is different, write it in parentheses next to your answer in (h).

3. The boxplots below summarize the Nielsen ratings for one summer week in 1989.

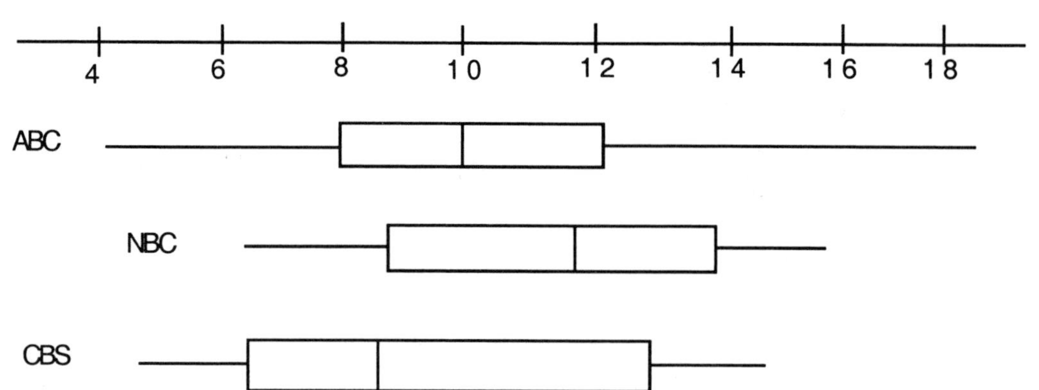

a. If winning the ratings for the week meant you had to have the highest rated show, which network would have won the ratings for that week? *ABC* ***(3)***

b. What network had the lowest rated show? *ABC* ***(3)***

c. What network had the highest median? *NBC* ***(3)***

d. If winning the ratings for the week meant you had to have the highest third quartile, which network would have won the ratings for that week? *NBC* ***(3)***

e. Which network do think won the ratings battle for the week? *NBC* ***(1)***
Explain. ***(6)***

Answer should include one or more of the following:
**highest median*
**highest third quartile*
best "worst" show
highest lowest quartile
* *More credit should be given for these answers*
An alternate answer could be ABC if belief that ratings are won with the best rated show.

f. Which network had the largest spread in its ratings? *ABC* ***(1)***
Explain how you made your choice. ***(6)***

Student gives range.
Accept NBC from any student who uses the difference between the two quartiles.

g. Officials at CBS were very unhappy when the ratings were announced. Explain why. ***(6)***

No matter what criteria are used: median, largest number etc., (except for lowest single rated show) CBS received the lowest score.

h. CBS decided to cancel the 4 shows with the lowest ratings. After cancelling the 4 shows their 5-number summary was: min: 6.9; first quartile: 9.1; median: 10.85; third quartile: 13.6; max: 14.6. Did cancelling the 4 worst shows help? <u>*No*</u> ***(1)*** Explain. ***(6)***

The ratings were improved but not enough. The median, Q3 and largest number are still less than NBC's, but they are now competitive. Q1 and lowest number are better than ABC's and NBC's.

4. a. Explain what the coach means by his statement? ***(6)***

The mean would be altered by the one very tall player and make the team appear taller than it really is. The median would not take the really tall player into account and make the team appear to be all short. The one tall player could make the team quite successful.

b. Is the mean or median the best measure of central tendency for this example? Justify your answer. ***(6)***

The student will have to make a choice and then defend it. One possible answer is "neither", since the difference between the mean and median will not be enough to sway the opponents one way or the other.

"Should we scare the opposition by announcing our mean height or lull them by announcing our median height?"

MATH *Connections* **I**
TURNING FACTS INTO IDEAS QUIZ
SECTIONS 1.6 - 1.7 (A)

Name ______________________________ Date ______________

1. Explain why the absolute value of seven (7) has the same value as the absolute value of negative seven (-7).

____

____

____

2. Find the value of each of the following:

a. $|5| =$ ____ b. $|-6+3| =$ ____

c. $|6-10| =$ ____ d. $|-3-(-5)| =$ ____

3. The Des Moines Pirates women's team has players with the following heights:
180 cm, 189 cm, 199 cm, 195 cm, 152 cm, 149 cm, 174 cm, 192 cm,
202 cm, 183 cm, 149 cm, 189 cm.

a. Find the mean for the players' heights to the nearest whole number. ________

b. Find the mean absolute deviation for the heights.

c. Explain what information you obtain about the heights of the players from the mean absolute deviation.

____

____

____

____

____

4. Determine a value for each of the following.

   a. 5.1^2 = ____________ b. $(\frac{2}{3})^2$ = ____________

   c. -3.2^2 = ____________ d. $(-0.032)^2$ = ____________

   e. $\sqrt{17.3}$ = ____________ f. $-\sqrt{473.1}$ = ____________

5. The heights of the players for the Des Moines Pirates are listed below.

   180 cm, 189 cm, 199 cm, 195 cm, 152 cm, 149 cm, 174 cm, 192 cm, 202 cm, 183 cm, 149 cm, 189 cm.

   a. Find the variance for the heights of the players. __________

   b. Find the standard deviation for the heights of the players. __________

   c. What information do you obtain about the heights of the players from the variance and the standard deviation?

   ____

   ____

   ____

   ____

   ____

   d. You can determine the value of $\sqrt{(505)^2}$ and $(\sqrt{505})^2$ without using a calculator. Explain why this is possible.

   ____

   ____

SOLUTION KEY AND SCORING GUIDE

MATH *Connections* **I**
TURNING FACTS INTO IDEAS QUIZ
SECTIONS 1.6 - 1.7 (A)

1. Explain why the absolute value of seven (7) has the same value as the absolute value of negative seven (-7). ***(6)***

   The values are the same because the absolute value of a number is the distance the number is from zero. Distance is always expressed as a positive number.

2. Find the value of each of the following. ***(5 each)***

   a. $|5| =$ <u>5</u> b. $|-6+3| =$ <u>3</u>

   c. $|6-10| =$ <u>4</u> d. $|-3-(-5)| =$ <u>2</u>

3. The Des Moines Pirates women's team has players with the following heights:
   180 cm, 189 cm, 199 cm, 195 cm, 152 cm, 149 cm, 174 cm, 192 cm, 202 cm, 183 cm, 149 cm, 189 cm.

   a. Find the mean for the players' heights to the nearest whole number. <u>*179 cm*</u> ***(5)***

   b. Find the mean absolute deviation for the heights. <u>*15.8*</u> ***(5)***

   The deviations are: 1, 10, 20, 16, 27, 30, 5, 13, 23, 4, 30, 10

   c. Explain what information you obtain about the heights of the players from the mean absolute deviation. ***(6)***

   The mean absolute variation is the average (mean) of the variations from the mean of the heights (179). In this case the mean absolute variation tells us that the spread of the data is relatively large. The team members differ in height by a significant amount.

4. Determine a value for each of the following. ***(5 each)***

a. 5.1^2 = *26.01*

b. $(\frac{2}{3})^2$ = *4/9*

c. -3.2^2 = *- 10.24*

d. $(-0.032)^2$ = *0.001024*

e. $\sqrt{17.3}$ = *4.16*

f. $-\sqrt{473.1}$ = *- 21.75*

5. The heights of the players for the Des Moines Pirates are listed below.

180 cm, 1898 cm, 199 cm, 195 cm, 152 cm, 149 cm, 174 cm, 192 cm, 202 cm, 183 cm, 149 cm, 1889 cm.

(Use the mean from problem 3.)

a. Find the variance for the heights f the players. *343.75* ***(8)***

The squares of each deviation are: 1, 100, 400, 256, 729, 900, 25, 169, 529, 16, 900, 100

b. Find the standard deviation for the heights of the players. *18.54* ***(8)***

c. What information do you obtain about the heights of the players from the variance and the standard deviation? ***(6)***

The spread of the data is significant. There are relatively tall and short players on the team. The team is not all the same height.

d. You can determine the value of $\sqrt{(505)^2}$ and $(\sqrt{505})^2$ without using a calculator. Explain why this is possible. ***(6)***

The answer is 505 in both examples. Squaring a number and taking the square root of a number are inverse operations which cancel each other out.

MATH *Connections* **I**
TURNING FACTS INTO IDEAS TEST (A)
CHAPTER 1: PART A

Name ______________________________ Date ____________________

Use the table below to answer the questions in this section.

AGES OF U.S. PRESIDENTS AT THEIR DEATH
The table below lists the presidents of the United States and the ages at which they died.

Washington	67	Fillmore	74	Roosevelt	60
Adams	90	Pierce	64	Taft	72
Jefferson	83	Buchanan	77	Wilson	67
Madison	85	Lincoln	56	Harding	57
Monroe	73	Johnson	66	Coolidge	60
Adams	80	Grant	63	Hoover	90
Jackson	78	Hayes	70	Roosevelt	63
Van Buren	79	Garfield	49	Truman	88
Harrison	68	Arthur	57	Eisenhower	78
Tyler	71	Cleveland	71	Kennedy	46
Polk	53	Harrison	67	Johnson	64
Taylor	65	McKinley	58		

1. Use your calculator to enter in the LISTS the ages of the presidents at their death.

2. Display the 1-variable STAT menu and determine the following:

   $\bar{x}$ ________
   Σx ________
   n ________

3. Explain the relationship among $\bar{x}$, Σx and n.

   ____

   ____

   ____

4. Sort the data that is stored in the calculator from least to greatest.
   What is the first number ($L_1(1)$) in your sort? _____
   What is the twelfth number ($L_1(12)$) in your sort? _____

5. Use the sorted data to find the median and mode(s).

   median ______
   mode ______

6. Use the sorted data to construct a stem-and-leaf plot:

Explain the meaning of 7|1 in this stem-and-leaf plot.

____

____

7. a. Use the stem and leaf plot from 6. above or your list to name the elements of the five number survey:

Least Value ______
Greatest Value ______
Median ______
First Quartile ______
Third Quartile ______

Explain what is meant by the First Quartile.

____

____

b. Use your calculator to construct the box plot for this data. Copy the boxplot onto your test paper below. Use the TRACE feature on your calculator to confirm your answers to 7 a. Correct any differences by writing the differences in parentheses next to your original answers.

8. If you were limited to using one data display to represent this data, which one would you choose, the stem-and-leaf or boxplot? List at least two reasons for your answer.

____

____

____

____

____

MATH *Connections* **I**
TURNING FACTS INTO IDEAS TEST
CHAPTER 1: PART B

Name ____ Date ______________

1. The boxplots below display the number of miles per gallon (mpg) for compact and medium sized cars produced in the United States by manufacturers A, B, and C.

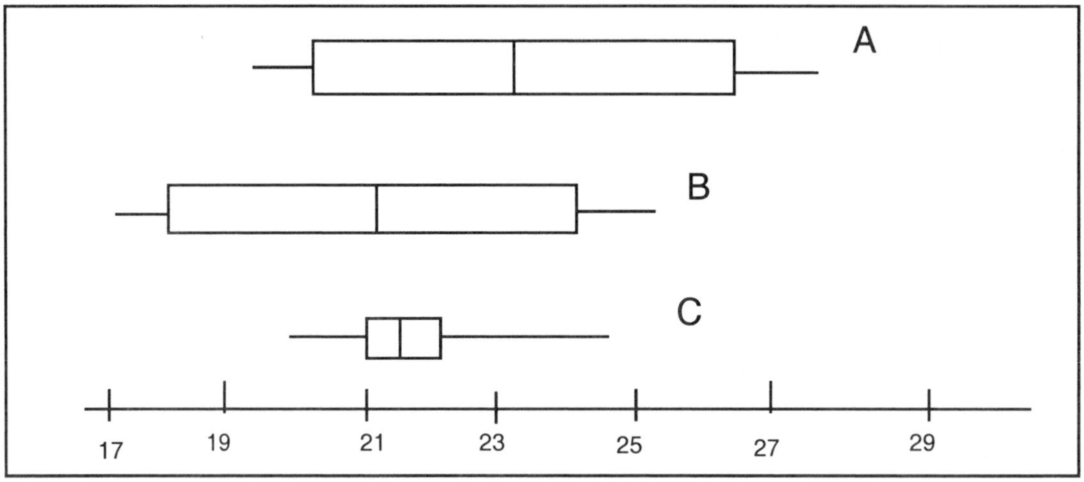

a. Name the company with the model with the highest mpg. ______

b. Name the company with the model with the least mpg. ______

c. Name the company with the least spread. _______

d. Suppose you work with manufacturer B and you want to improve your mpg compared to A and C. Should you put extra effort into improving your cars with the most mpg, improving your cars with the least mpg or should you spread your effort over all the cars? Explain your answer.

____

____

____

____

e. Jose wishes to buy a car and would like the mpg to be above 21. If Jose wanted to be reasonably sure to purchase a car with mpg of 21 or better, would he buy from manufacturers A, B, or C? Explain your answer.

____

____

____

____

2. The following histogram shows the cost of food per day in U.S. cities with very high travel costs and the number of cities within a range of costs.

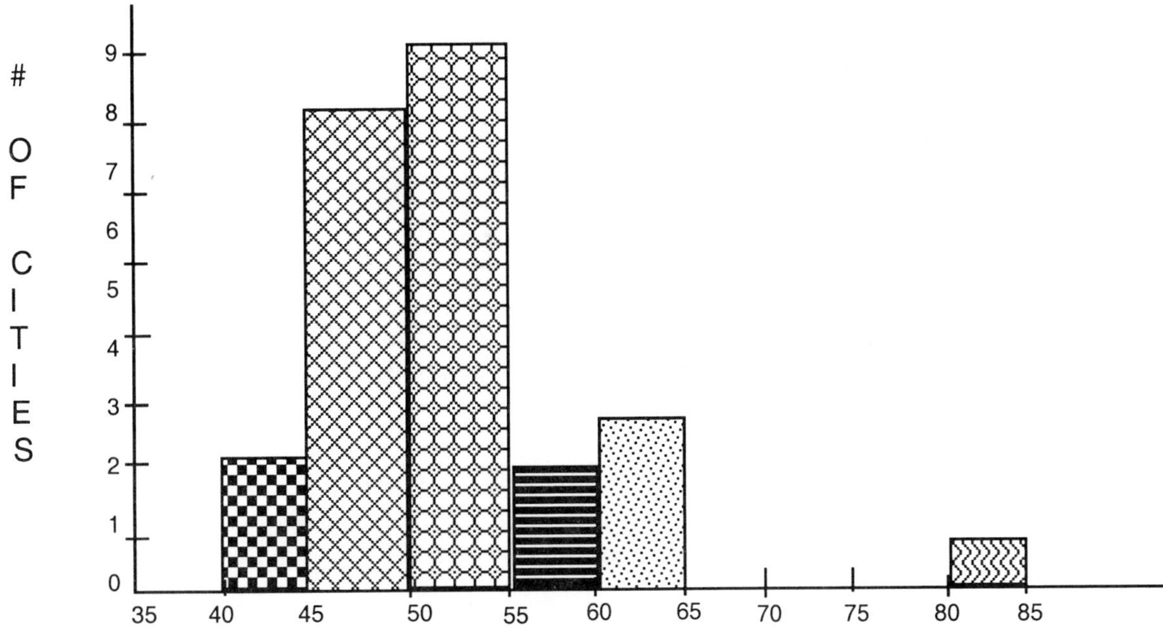

a. How many cities were involved in this data set? _____

b. The food cost for Hartford is $52.30. How does this compare to the median?

____

____

____

____

c. New York City is represented by the bar to the far right. What can you say about the daily cost of food for a traveler to New York City?

____

____

d. If you were travelling to six of these cities, what would be a reasonable daily amount of money to budget for food costs? Explain your answer.

____

____

____

e. List two questions you have about this data that you would not be able to answer from the histogram.

____

____

____

____

3. Jan's height is 70 in. That is the mean height of the players on her volleyball team. Bill is 68 in., Edna is 66 in., Karl is 72 in., Peter is 78 in. and Kara is 66 in.
   a. Make a balance picture of the six heights.

   b. Find the mean absolute variation of these heights. _________

4. The following data deals with sneaker prices of popular brands of sneakers sold at Karl's Discount Store:

SNEAKER PRICES	
Air Jordan by Nike	$125
Air Macs by Nike	$110
Air 180 by Nike	$125
Air Tech Challenge by Nike	$85
Asics Running Shoe`	$85
Blacktop by Reebok	$70
Bo Jackson by Nike	$110
Converse All-Star	$30
Ex-O-Fit by Nike	$60
Flight by Nike	$80
Force by Nike	$75
Pump by Reebok	$125
Puma	$80

Find the variance and standard deviation for the sneaker prices.

variance: ____________ standard deviation: ______________

SOLUTION KEY AND SCORING GUIDE
MATH *Connections* **I**
TURNING FACTS INTO IDEAS TEST (A)
CHAPTER 1: PART A

Use the table below to answer the questions in this section.

AGES OF U.S. PRESIDENTS AT THEIR DEATH
The table below lists the presidents of the United States and the ages at which they died.

Washington	67	Fillmore	74	Roosevelt	60
Adams	90	Pierce	64	Taft	72
Jefferson	83	Buchanan	77	Wilson	67
Madison	85	Lincoln	56	Harding	57
Monroe	73	Johnson	66	Coolidge	60
Adams	80	Grant	63	Hoover	90
Jackson	78	Hayes	70	Roosevelt	63
Van Buren	79	Garfield	49	Truman	88
Harrison	68	Arthur	57	Eisenhower	78
Tyler	71	Cleveland	71	Kennedy	46
Polk	53	Harrison	67	Johnson	64
Taylor	65	McKinley	58		

1. Enter the ages of the presidents at their death by using the EDIT option on the STAT menu on a graphing calculator.
2. Display the 1-variable STAT menu and determine the following: ***(2 each)***

   $\bar{x}$ *68.8*

   Σx *2 409*

   n *35*
3. Explain the relationship among $\bar{x}$, Σx and n. (6) ***(4)***

   To find $\bar{x}$, the mean, we need to find the sum of all the ages, Σx, and divide the sum by n, the number of presidents.
4. Sort the data that is stored in the calculator from least to greatest.
   What is the first number ($L_1(1)$) in your sort? *46* What is the twelfth number ($L_1(12)$) in your sort? *64* ***(2 each)***
5. Use the sorted data to find the median and mode(s). ***(2 each)***

   median *67*

   mode *67*
6. Use the sorted data to construct a stem-and-leaf plot: ***(5)***

4	**6 9**
5	**3 6 6 7 8**
6	**0 0 3 3 4 4 5 6 7 7 7 8**
7	**0 1 1 2 3 4 7 8 8 9**
8	**0 3 5 8**
9	**0 0**

Explain the meaning of 7|1 in this stem-and-leaf plot. ***(2)***

7|1 means a president died at age 71 years.

7. a. Use the stem and leaf plot from 6. above or your list to name the elements of the five number survey: ***(1 each)***

Least Value	*46*
Greatest Value	*90*
Median	*67*
First Quartile	*60*
Third Quartile	*78*

Explain what is meant by the First Quartile. ***(6)***
The First Quartile is the number which has one-fourth of the data items below it.

b. Use your calculator to construct the box plot for this data. Copy the boxplot onto your test paper below. Use the TRACE feature on your calculator to confirm your answers to 7 a. Correct any differences. ***(4)***

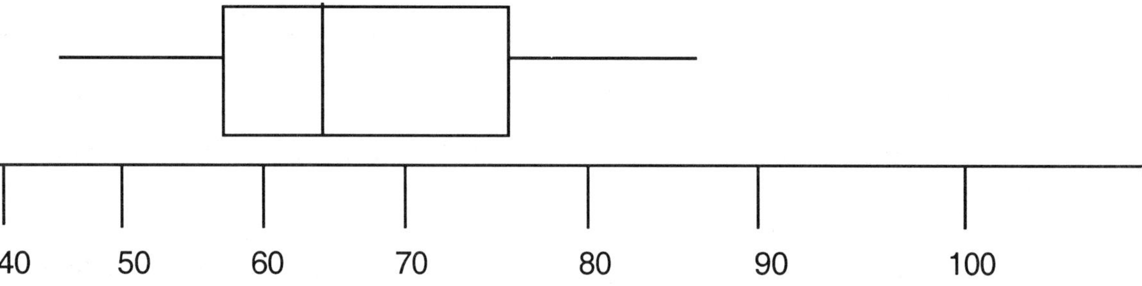

40 50 60 70 80 90 100

8. If you were limited to using one data display to represent this data, which one would you choose, the stem-and-leaf or boxplot? List at least two reasons for your answer. ***(6)***

A rich answer should include some of the following:
For those choosing the stem-and-leaf

a. it shows the original data
b. it can be rotated and viewed as a histogram
c. mean can be computed
d. mode can be computed
e. no big gaps can be seen immediately

For those choosing the box plot:

a. it proves a quick visual summary
b. large and smallest data values can be read quickly
c. Median, Q1 and Q3 can be read quickly
d. the behavior of middle 50% is easily seen
e. large and small data items in comparison to rest of data are easily seen

Any two of the above are fine
Student with only one answer should not receive many points. Student with more than two answers should receive maximum credit

SOLUTION KEY AND SCORING GUIDE
MATH *Connections* I
TURNING FACTS INTO IDEAS TEST
CHAPTER 1: PART B

1. The boxplots below display the number of miles per gallon (mpg) for compact and medium sized cars produced in the United States by manufacturers A, B, and C.

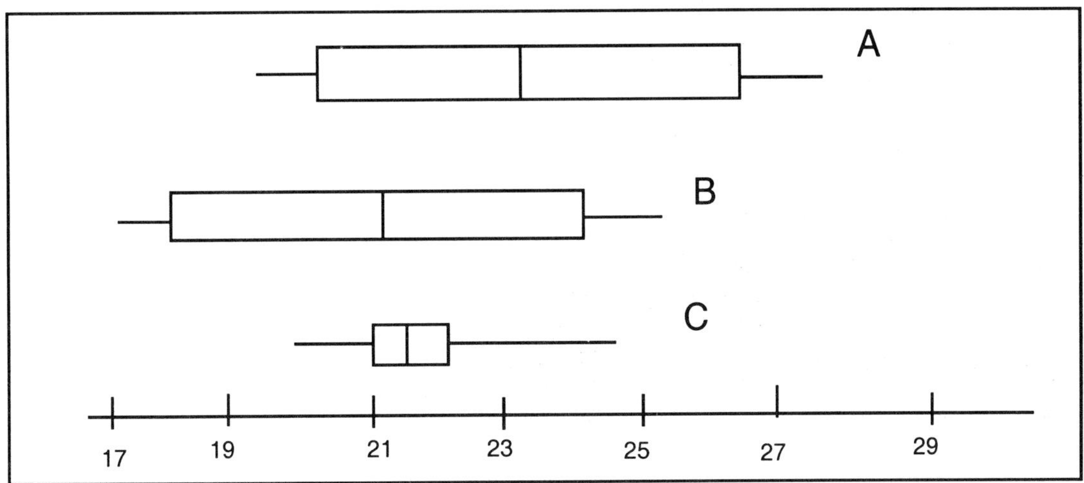

a. Name the company with the model with the highest mpg. *A* **(2)**

b. Name the company with the model with the least mpg. *B* **(2)**

c. Name the company with the least spread. *C* **(2)**

d. Suppose you work with manufacturer B and you want to improve your mpg compared to A and C. Should you put extra effort into improving your cars with the most mpg, improving your cars with the least mpg or should you spread your effort over all the cars? Explain your answer. ***(6)***

The least because Q1, the median and Q3 reflect location when the data is ordered. By mproving the worst, the lowest number, Q1 and the median will be increased. Give more credit for the following commentary in addition to the above: B's top is already better than C's and is competitive with A's top 50%. An increase in the median accomplished by improving the cars with the lowest mpg will make its top 50% competitive with C's top 50% and still better than C's top 25%.

e. Jose wishes to buy a car and would like the mpg to be above 21. If Jose wanted to be reasonably sure to purchase a car with mpg of 21 or better, would he buy from manufacturers A, B, or C? Explain your answer. ***(6)***

C. *75% of C's cars have mpg 21 or better. Only 50% of B's do.*

2. The following histogram shows the cost of food per day in U.S. cities with very high travel costs and the number of cities within a range of costs.

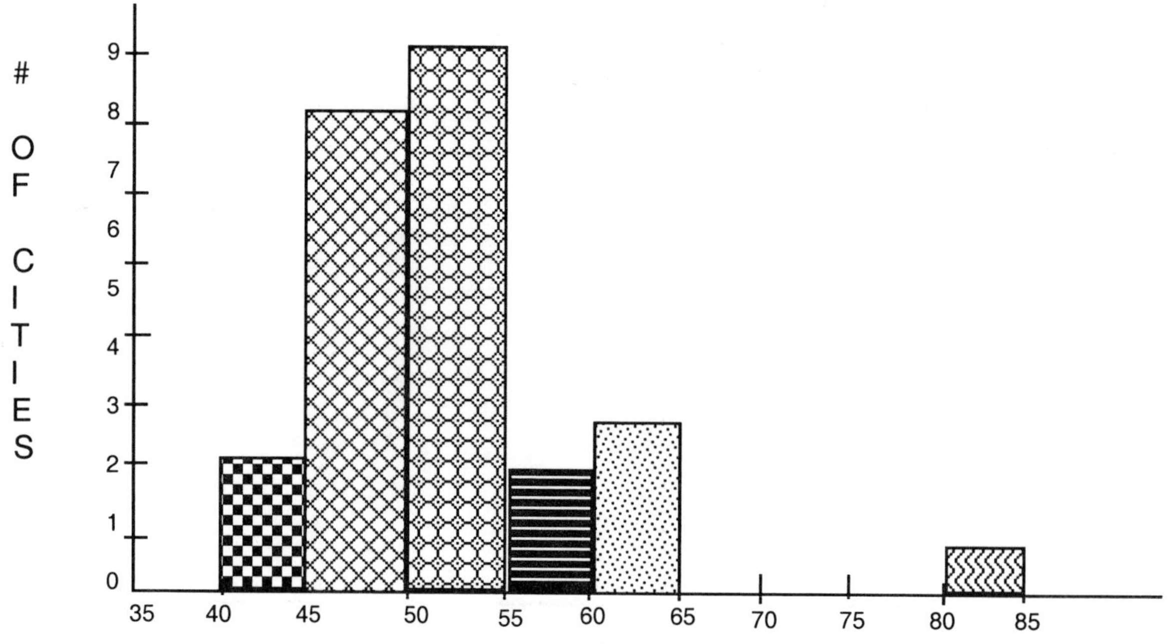

a. How many cities were involved in this data set? <u>25</u> ***(2)***

b. The food cost for Hartford is $52.30. How does this compare to the median? ***(4)***
The median cost is about $50 to $55 and $52.50 is in that interval. It is not possible to tell if Hartford is above or below the median.

c. New York City is represented by the bar to the far right. What can you say about the daily cost of food for a traveler to New York City? ***(4)***
Food costs about $80 a day and that is much more compared to all the other cities in the study.

d. If you were travelling to six of these cities, what would be a reasonable daily amount of money to budget for food costs? Explain your answer. ***(6)***
$55. More than half of the cities charge less than this amount and a number were considerably lower.
Answers could vary for this problem. Students should use good reasoning to defend answer. Accept any answer that is more than $55 and less than $85

e. List two questions you have about this data that you would not be able to answer from the histogram. ***(4)***
Some possible responses are:
i. Which two cities are least expensive?
ii. Which cities had food expenses that were more than the median.
iii. Which cities were studied ?
iv. Were the cities studied the only ones with high travel costs?

3. Jan's height is 70 in. That is the mean height of the players on her volleyball team. Bill is 68 in., Edna is 66 in., Karl is 72 in., Peter is 78 in. and Kara is 66 in.

   a. Make a balance picture of the six heights. ***(3)***

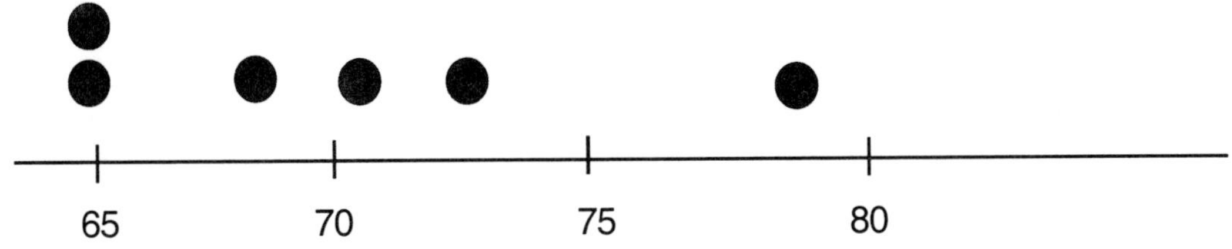

   b. Find the mean absolute variation of these heights. __*3.33*__ ***(3)***

4. The following data deals with sneaker prices of popular brands of sneakers sold at Karl's Discount Store:

SNEAKER PRICES	
Air Jordan by Nike	$125
Air Macs by Nike	$110
Air 180 by Nike	$125
Air Tech Challenge by Nike	$85
Asics Running Shoe`	$85
Blacktop by Reebok	$70
Bo Jackson by Nike	$110
Converse All-Star	$30
Ex-O-Fit by Nike	$60
Flight by Nike	$80
Force by Nike	$75
Pump by Reebok	$125
Puma	$80

Find the variance and standard deviation for the sneaker prices. ***(5 each)***

variance: __*757.15*__ standard deviation: __*7.63*__

MATH *Connections* **I**
WELCOME TO ALGEBRA QUIZ
SECTIONS 2.1 - 2.2 (A)

Name ______________________________________ Date ________________

1. The Central Connecticut Electric Company charges a basic fee of $8.50 a month and, in addition, $.09 for each kiloWatt-hour that is used during the month. If *u* represents the number of kiloWatts used in a month, then the total charge, *T*, can be determined by using the formula:

$$T = \$8.50 + \$0.09 * u$$

   a. Identify the variables and constants in the algebraic sentence: $T = 8.5 + 0.09 * u$ and explain why each is a variable or a constant.

   ____

   ____

   ____

   ____

   b. Name one advantage and one disadvantage of using "T" and "u" as abbreviations in this problem.

   c. Find the total cost of electricity for the months listed below:

   January kiloWatts used: 280 Total cost: ____________

   August kiloWatts used: 477 Total cost ____________

   d. Why do you think that the kiloWatt use was higher in August than January?

   ____

   ____

   ____

   ____

2. To find the cost of ordering pairs of jeans from a catalogue the following equation is used:

$$C = \$21.98 * N + \$5.50$$

where C is the total cost, N is the number of jeans to be ordered, \$21.98 represents the cost of each pair of jeans, and \$5.50 is the shipping cost per order. If Tom and his brothers decide to order 5 pair of jeans, what will be the total cost?

____

3. Binh gathered the following data from an experiment (temperature after t minutes) in science class:

Time	Temperature
3 min.	74°
6 min.	76°
9 min.	78°
12 min.	80°
15min.	82°

a. Explain the pattern that is shown in Binh's experiment.

____

____

b. Using this pattern, predict the temperature after 21 minutes. ________

4. In her math class, Elsa was given the following information about a pattern:

Input(n)	Output(A)
7	12
8	13
9	14
10	15

a. If the input is represented by the variable, n, write an algebraic sentence to determine the output, A. ______________________

b. If the input, n, is 20 what is the output, A? __________________

5. Chirps of crickets can tell us the Fahrenheit temperature. To find the temperature you use the following algorithm:

   The temperature in Fahrenheit degrees is determined by adding 40° to one-half the number of chirps per minute.

   a. Select a variable for the temperature _____ and a variable for the number of chirps per minute _____.

   b. Write an equation that shows the relationship between the temperature and the number of chirps per minute. ________________________

6. Use the equation you wrote in 5 to:

   a. Find the temperature if the number of chirps per minute is 76.

   b. Find the number of chirps per minute if the temperature is 60°.

7. Write expressions for each of the following. State what each variable represents for each problem.

   a. Tori has dimes and quarters in her bank. Write an expression that Tori can use to find the value of the money in her bank.

   b. A parcel shipping service charges a $6.00 fee plus $1.50 for each pound. Write an expression for determining the cost for shipping a parcel through this service.

c. In Australian Rules Football, you receive 6 points for each ball kicked between the middle posts (goal) and 1 point for each ball kicked between the outside posts (goal). Write an expression for determining a team's final score in an Aussie football match.

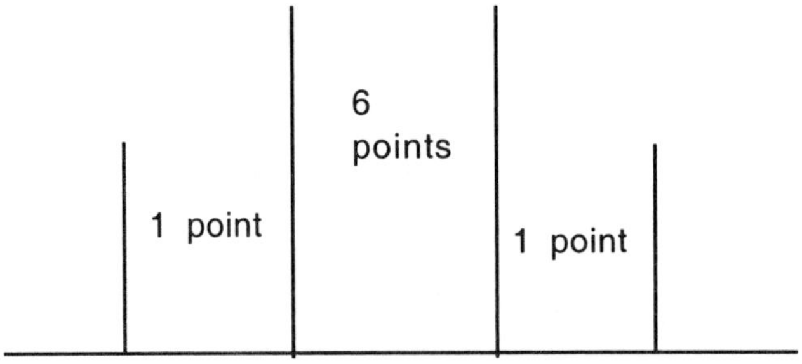

6
points
1 point
1 point

SOLUTION KEY AND SCORING GUIDE

MATH *Connections* **I**
WELCOME TO ALGEBRA QUIZ
SECTIONS 2.1 - 2.2 (A)

1. The Central Connecticut Electric Company charges a basic fee of $8.50 a month and, in addition, $.09 for each kiloWatt-hour that is used during the month. If *u* represents the number of kiloWatts used in a month, then the total charge, *T*, can be determined by using the formula:

   $T = \$8.50 + \$0.09 * u$

   a. Identify the variables and constants in the algebraic sentence: $T = 8.5 + 0.09 * u$ and explain why each is a variable or constant. ***(12)***

   T and u are variables as their values can change or vary depending on how many kiloWatt-hours a customer uses during a month.
   8.5 and 0.09 are constants as their values do not change as the charge per kiloWatt-hour, and the monthly basic fee are the same each month.

   b. Name one advantage and one disadvantage of using "T" and "u" as abbreviations in this problem. ***(6)***

   One advantage is that it saves space. A second would be that the use of T and u are related to the context of the problem. A disadvantage is that the reader would not know what T and u represented unless they knew the context.

   c. Find the total cost of electricity for the months listed below: ***(4 each)***

January	kiloWatts used: 280	Total cost: *$33.70*
August	kiloWatts used: 477	Total cost *$51.43*

   d. Why do you think that the kiloWatt use was higher in August than January? ***(6)***

   Possible answers might be that an air conditioner may be operated in the summer as well as a pool filter.

2. To find the cost of ordering pairs of jeans from a catalogue the following equation is used:

   $C = \$21.98 * N + \$5.50$

   where C is the total cost, N is the number of jeans to be ordered, $21.98 represents the cost of each pair of jeans and $5.50 is the shipping cost per order. If Tom and his brothers decide to order 5 pair of jeans, what will be the total cost?
   $115.40 ***(6)***

3. Binh gathered the following data from an experiment (temperature after *t* minutes) in science class:

Time	Temperature
3 min	74°
6 min	76°
9 min	78°
12 min	80°
15min	82°

a. Explain the pattern that is shown in Binh's experiment. ***(6)***

The pattern is that every three minutes the temperature rises 2 degrees.

b. Using this pattern, predict the temperature after 21 minutes. _*86°*_ ***(6)***

4. In her math class, Elsa was given the following information about a pattern:

Input(n)	Output(A)
7	12
8	13
9	14
10	15

a. If the input is represented by the variable, *n*, write an algebraic sentence to determine the output, *A*. $A = n + 5$ ***(6)***

b. If the input, *n*, is 20 what is the output, *A*? _*25*_ ***(6)***

5. Chirps of crickets can tell us the Fahrenheit temperature. To find the temperature you use the following algorithm:

The temperature in Fahrenheit degrees is determined by adding 40° to one-half the number of chirps per minute.

a. Select a variable for the temperature _*answers will vary*_ and a variable for the number of chirps per minute _*answers will vary*_. ***(6)***

b. Write an equation that shows the relationship between the temperature and the number of chirps per minute. $T = (40 + 1/2\ C)^\circ$ ***(6)***

6. Use the equation you wrote in 5 to: ***(5 each)***

   a. Find the temperature if the number of chirps per minute is 76.

   $T = (40 + 1/2 * 76)° = 78°$

   b. Find the number of chirps per minute if the temperature is 60°.

   $60 = (40 + 1/2 * C)°$; $C = 40$ *chirps*

7. Write expressions for each of the following. State what each variable represents for each problem.

   a. Tori has dimes and quarters in her bank. Write an expression that Tori can use to find the value of the money in her bank. ***(6)***

   Choose D to represent the number of dimes and Q to represent the number of quarters
   $T = \$.10 * D + \$.25 * Q$

   b. A parcel shipping service charges a $6.00 fee plus $1.50 for each pound. Write an expression for determining the cost for shipping a parcel through this service. ***(6)***

   Choose T to represent the cost and P to represent the weight in pounds
   $T = \$6 + \$1.50 * P$

   c. In Australian Rules Football, you receive 6 points for each ball kicked between the middle posts (goal) and 1 point for each ball kicked between the outside posts (goal). Write an expression for determining a team's final score in an Aussie football match. ***(4)***

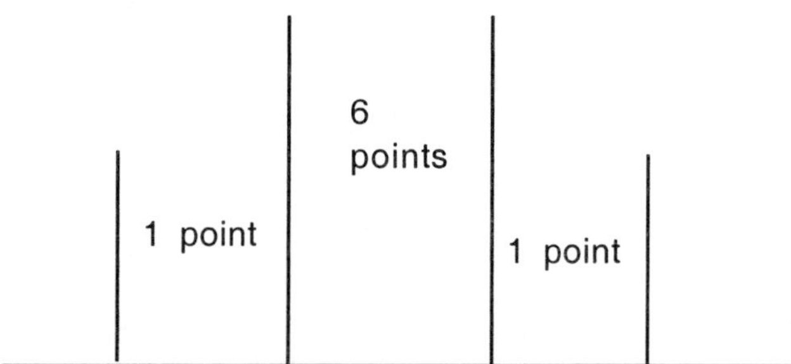

Choose T to represent the final score, G to represent the goals and E to represent the kicks through to outside posts.

$T = 6 * G + 1 * E$

MATH *Connections* **I**
WELCOME TO ALGEBRA QUIZ
SECTIONS 2.3 - 2.4 (A)

Name ________________________________ Date ______________

1. Joanna takes the following quiz on order of operations.
   - Correct the quiz. Next to the answer put an "x" if wrong, a "C" if correct.
   - If you believe the answer is wrong, write the correct answer in the blank.

a. $16 \div 4 * 2 + 6 - 3$ answer: 5 ________

b. $5 + 3(7 - 2)$ answer: 20 ________

c. $6 \div 3 + 3 * 3$ answer: 11 ________

d. $20 - \frac{18}{3} * 2$ answer: 28 ________

e. $12 - 8 + 10 - 14 \div 14$ answer: 13 ________

2. Explain the **order of operations** rules.

____

____

____

____

3. Are the following commutative activities? (Answer yes or no)

a. Put on socks ________
Put on shoes

b. Unlock door ________
Open door

c. Mow lawn ________
Weed garden

d. Buy flour ________
Buy sugar

4. Marlene took the following true-false quiz on the Commutative and Associative Laws. Circle the problems with which you agree with Marlene's answer.

a. $6 + 12 = 12 + 6$ True

b. $6 - 12 = 12 - 6$ True

c. $(4 + 7) + 10 = 4 + (7 + 10)$ False

d. $(4 * 7) * 10 = 4 * (7 * 10)$ True

e $5 * 8 = 8 * 5$ True

f. $(4 - 7) - 10 = 4 - (7 - 10)$ True

g. $12 \div 6 = 6 \div 12$ False

Write conclusions about commutativity and associativity with regard to addition, subtraction, multiplication, and division.

____

____

____

____

____

5. Apply the Distributive Law to each of the following and rewrite the expression without parentheses:

a. $6 * (c + 5)$ ______________________

b. $(12 - t) * 3$ ______________________

c. $5(x - 2 + y)$ ______________________

d. $(y + 2) * 4$ ______________________

6. Without using a calculator, demonstrate how the Commutative, Associative and Distributive Laws can assist in making each computation easier:

a. $5 * 99$

b. $12 * 11$

c. $12 * (3\frac{1}{4})$

d. $5.25 * 20$

e. $(1200 + 71) + (800 + 9)$

7. Write the inverse operation or inverse activity for each of the following:

   a. Putting on a glove ______________________________

   b. Digging a hole ______________________________

   c. Adding 3 ______________________________

   d. Dividing by 0.1 ______________________________

8. Sometimes equations have no solution or infinitely many solutions.

   a. Write an example of an equation that has no solution:

   b. Write an example of an equation that has infinitely many solutions:

9. Explain the algorithm you would use to find the solution for the equation:

$$50 = 7 * t + 1$$

10. Mark and Samantha are in charge of designing the first part of today's quiz. They decided to think up some equations and find the solutions. They wrote eight equations on separate pieces of paper and wrote the answers on other separate pieces of paper. Just as they were ready to hand the equations and answers to their teacher, a wind blew them all on the floor and scrambled them. Your job is to put the mess back together.
Match the equation with its solution.

1. $28 = 5a + 3$

b. 7

a. 3

2. $40 = 7a + 40$

3. $13 = 2a + 1$

c. 4

4. $18 = 4a + 2$

d. 0

e. 5

6. $21 = 7a + 0$

f. 1

5. $29 = 8a + 13$

7. $16 = 10a + 6$

g. 2

8. $54 = 7a + 5$

h. 6

1. _________ 2. _________ 3. _________ 4. _________

5. _________ 6. _________ 7. _________ 8. _________

11. For each one the following (a - c):
    1. Write one sentence stating what you are looking for in the problem
    2. Assign an appropriate variable to what you are looking for
    3. Write an equation to describe the problem
    4. Solve the equation

a. Myra has $20.25. She and her brother have $40.65 together. Find the amount of money her brother has.

b. Al jogs at 5 mph. He jogs 10 miles every day. Yesterday he stopped to chat with his girl friend for 1 hour and 15 minutes. What was the total time it took Al to complete his 10 mile jog yesterday?

c. Tawhnee buys 40 feet of lumber at 89¢ a foot. She used 33 feet. How much did she pay for the lumber she did not use?

SOLUTION KEY AND SCORING GUIDE

MATH *Connections* **I**
WELCOME TO ALGEBRA QUIZ
SECTIONS 2.3 - 2.4 (A)

1. Joanna takes the following quiz on order of operations.
   - Correct the quiz. Next to the answer put an "X" if wrong, a "C" if correct.
   - If you believe the answer is wrong, write the correct answer in the blank.

   (2 each)

   a. $16 \div 4 * 2 + 6 - 3$ answer: 5 *X* *11*

   b. $5 + 3(7 - 2)$ answer: 20 *C* ____

   c. $6 \div 3 + 3 * 3$ answer: 11 *C* ____

   d. $20 - \frac{18}{3} * 2$ answer: 28 *X* *8*

   e. $12 - 8 + 10 - 14 \div 14$ answer: 13 *C* ____

2. Explain the **order of operations** rules. ***(6)***

   Student should include : parentheses first, multiplication and division from left to right, addition and subtraction from left to right

3. Are the following commutative activities? (Answer yes or no.) ***(2 each)***

   a. Put on socks *no*
   Put on shoes

   b. Unlock door *no*
   Open door

   c. Mow lawn *yes*
   Weed garden

   d. Buy flour *yes*
   Buy sugar

4. Marlene took the following true-false quiz on the Commutative and Associative Laws. Circle the problems with which you agree with Marlene's answer. ***(7)***

   (a. $6 + 12 = 12 + 6$ True) [circled]

b. 6 – 12 = 12 – 6 True

c. (4 + 7) + 10 = 4 + (7 + 10) False

d. (4 * 7) * 10 = 4 * (7 * 10) True

e 5 * 8 = 8 * 5 True

f. (4 – 7) – 10 = 4 – (7 – 10) True

g. 12 ÷ 6 = 6 ÷ 12 False

Write what conclusions you can make about commutativity and associativity with regard to addition, subtraction, multiplication and division. ***(6)***

A rich answer would include the following:

- *The operations of addition and multiplication are both commutative and associative*
- *The operations of subtraction and division are not commutative or associative.*

5. Apply the Distributive Law to each and rewrite as an expression without parentheses:

(2 each)

a. 6 * (c + 5) *6c + 30*

b. (12 - t) * 3 *36 - 3t*

c. 5(x - 2 + y) *5x - 10 + 5y*

d. (y + 2)4 *4y + 8*

6. Without using a calculator, demonstrate how the Commutative, Associative and Distributive Laws can assist in making each computation easier: ***(2 each)***

a. 5 * 99

5 (100 – 1) = 500 – 5 = 495

b. 12 * 11

*12 *(10 + 1) = 120 + 12 = 132*

c. $12 * 3\frac{1}{4}$

$12 \quad (3 + \frac{1}{4}) = 36 + 3 = 39$

d. $5.25 * 20$

$(5 + \frac{1}{4}) * 20 = 100 + 5 = 105$

e. $(1200 + 71) + (800 + 9)$

$(1200 + 800) + (71 + 9) = 2000 + 80 = 2080$

7. Write the inverse operation or inverse activity for each of the following: ***(2 each)***

   a. Putting on a glove <u>*taking off a glove*</u>

   b. Digging a hole <u>*filling a hole*</u>

   c. Adding 3 <u>*subtracting 3*</u>

   d. Dividing by 0.1 <u>*multiplying by 0.1*</u>

8. Sometimes equations have no solution or infinitely many solutions.

   a. Write an example of an equation that has no solution: ***(3 each)***

   An example would be: $a + 1 = a$

   b. Write an example of an equation that has infinitely many solutions:

   An example would be: $a + 1 = a + 1$

9. Explain the algorithm you would use to find the solution for the equation:

$$50 = 7t + 1 \qquad \textbf{(6)}$$

The first step is to subtract one from each side.
The second step is to divide each side by seven.

10. Mark and Samantha are in charge of designing the first part of today's quiz. They decided to think up some equations and find the solutions. They wrote eight equations on separate pieces of paper and wrote the answers on separate pieces of paper. Just as they were ready to hand the equations and answers to their teacher, a wind blew them all on the floor and scrambled them. Your job is to put the mess back together.
Match the equation with its solution. ***(16)***

1. $28 = 5a + 3$ b. 7

a. 3 2. $40 = 7a + 40$

3. $13 = 2a + 1$

c. 4

4. $18 = 4a + 2$ d. 0

e. 5 6. $21 = 7a + 0$

f. 1

5. $29 = 8a + 13$

7. $16 = 10a + 6$ g. 2 8. $54 = 7a + 5$ h. 6

1. *e* 2. *d* 3. *h* 4. *c*

5. *g* 6. *a* 7. *f* 8. *b*

11. For each of the following problems (a - c): ***(3 each)***
    1. Write one sentence stating what you are looking for in the problem
    2. Assign an appropriate variable to what you are looking for
    3. Write an equation to describe the problem
    4. Solve the equation

a. Myra has \$20.25 . She and her brother have \$40.65 together. Find the amount of money her brother has.

1. *You are to find how much money her brother has.*
2. *Let B be the amount of money the brother has*
3. $B + 20.25 = 40.65$
4. $B = \$20.40$

b. Al jogs at 5 mph. He jogs 10 miles every day. Yesterday he stopped to chat with his girl friend for 1 hour and 15 minutes. What was the total time it took Al to complete his 10 mile jog yesterday?

1. *You are asked the total time Al took to jog.*
2. *Let T be the total time that Al jogged.*
3. *T = (10 ÷ 5) + 1 1/4*
4. *T = 2 + 1 1/4 = 3 1/4 hours*

c. Tawhnee buys 40 feet of lumber at 89¢ a foot. She uses 33 feet. How much did she pay for the part of the lumber she did not use?

1. *You are asked to find the cost of the unused lumber.*
2. *Let C be the cost of the unused lumber.*
3. *C = (40 - 33) $0.89 or 7 $0.89*
4. *C = $6.23*

MATH *Connections* **I**
WELCOME TO ALGEBRA QUIZ
SECTIONS 2.5 - 2.7 (A)

Name ____ Date________________

1. Julius was reviewing for this quiz and has forgotten some important information about exponents. He asks you to assist him. He asks you the following questions. Write a complete explanation for each of Julius' questions.

   a. What does 3^5 mean?

   b. I have $5 \cdot 5 \cdot 5 \cdot 2 \cdot 2 \cdot 2 \cdot 2$. How do I write the expression using exponents?

   c. Evaluate each of these expressions. Please explain each one to me.

   i. 2^6

   ii. 1.02^0

   iii. $3-0.2^4$

   iv. $(6 + 0.5)^2$

   v. 3^{-2}

   vi. $4 * 5^2$

   vii. $16 * 2^{-3}$

2. Write each number in scientific notation as a calculator would write it. For example, 132 would be written as 1.32E2 on your calculator.

   a. 80000000 ______________________________

   b. 0.037 ______________________________

   c. The amount the automobile industry spent on advertising in 1991 is $5,259,100,000. ______________________________

   d. The number of kilometers in a millimeter is 0.000001. _________

3. Write each number in standard notation:

   a. 4.3E4 ______________________________

   b. 5.1E – 4 ______________________________

   c. The population of Germany is 7.75E7. ______________________

   d. The price of French Fries at Wendy's is 8.9E ⁻1. ______________

4. Find each product:

   a. $5 * 10^3$

   b. $4 * 10^{-3}$

5. A formula for finding compound interest on a bank account is as follows:

$$T = P * (1 + i)^n$$

where T is the amount of money in the bank after n periods of compounding, P represents the amount of money on deposit before the interest is applied, and i represents the rate of interest.

- If you deposit $1000 at an interest rate of 4% compounded each year (annually) and leave it there for five years, how much money will you have in the bank after 5 years?

6. Use the laws of exponents to perform the following operations.

a. $x^2 * x^3$ ________ b. $2^6 * 2^3$ ________

c. $3x * 5x^4$ ________ d. $6a^2b * -2ab^2$ ________

e. $\frac{x^5}{x^2}$ ________ f. $\frac{11^7}{11^3}$ ________

g. $\frac{12x^6}{3x^2}$ ________ h. $\frac{-25x^4}{-5x^4}$ ________

i. $\frac{-27a^2b^4}{9a^3b^3}$ ________ j. $(x^2)^5$ ________

k. $(3x^5)^2$ ________ l. -5^2 ________

m. $(-2x^2y)^3$ ________

7. a. Evaluate each of the following.

i. $5 * 3^2$ ii. $4 + 6 * 3^2$

iii. $3 * 2^3 - 2 * 3^2$ iv. $5 + 2 * (7-3)^2$

b. Explain how exponentiation fits into the rules for order of operation.

____

____

SOLUTION KEY AND SCORING GUIDE

MATH *Connections* **I**
WELCOME TO ALGEBRA QUIZ
SECTIONS 2.5 - 2.7 (A)

1. Julius was reviewing for this quiz and has forgotten some important information about exponents. He asks you to assist him. He asks you the following questions. Write a complete explanation for each of Julius' questions.

   a. What does 3^5 mean? ***(3)***

   *3^5 means 3 *3 *3 *3 *3 which equals 243. The exponent tells how many times to use the base as a factor.*

   b. If I have *5 *5 *5 *2 *2 *2 *2,* how do I write the expression using exponents? ***(3)***

   Since an exponent indicates how many tImes the base is used as a factor, you count the number of 5's and conclude the exponent for 5 is 3. Then you count the number of 2's and conclude the exponent of 2 is 4. The expression will be:

   $5^3 * 2^4$

   c. Evaluate each of these expressions. Please explain each one to me.

   i. 2^6 ***(3)***

   *2^6 is the same as 2 *2 * 2 *2 *2 * 2 which is 64.*

   ii. 1.02^0 ***(3)***

   Raising any number, except 0, to the 0 power is 1, so $1.02^0 = 1$.

   iii. $3 - 0.2^4$ ***(3)***

   $3 - 0.2^4 = 3 - 2 * 2 * 2 * 2 = 3 - 0.0016 = 2.9984$

   iv. $(6+0.5)^2$ ***(3)***

   $(6+0.5)^2 = (6.5)^2 = 42.25$

   v. 3^{-2} ***(3)***

   $3^{-2} = \frac{1}{3^2} = \frac{1}{9}$

   vi. $4 * 5^2$ ***(3)***

   $4 * 5 = 4 * 5 * 5 = 100$

   vii. $16 * 2^2 = 16 * \frac{1}{8} = 2$ ***(3)***

2. Write each number in scientific notation as a calculator would write it. For example, 132 would be written as 1.32E3 on your calculator. ***(3 each)***

   a. 80000000 *8 E7*

   b. 0.037 *3.7 E⁻2*

   c. $5,259,100,000 , the amount the automobile industry spent on advertising in 1991. *5.2591 E 9*

   d. 0.000001, the number of kilometers in a millimeter. *1.0 E -6*

3. Write each number in standard notation: ***(3 each)***

   a. 4.3E4 *43000*

   b. 5.1E⁻4 *.00051*

   c. 7.75E7, the population of Germany *77,500,000*

   d. 8.9E ⁻1, the price of French Fries at Wendy's *$.89*

4. Find each product: ***(2 each)***

   a. $5 * 10^3$

   *5 * 1000 = 5000*

   b. $4 * 10^{-3}$

   *4 * .001 = .004*

5. A formula for finding compound interest on a bank account is as follows:

$$T = P * (1 + i)^n$$

where T is the amount of money in the bank after n periods of compounding, P represents the amount of money on deposit before the interest is applied, and i represents the rate of interest.

- If you deposit $1000 at an interest rate of 4% compounded each year and leave it there for five years, how much money will you have in the bank after 5 years? ***(5)***

$T = P * (1 + i)^n$
$T = 1000 * (1 + .04)^5$
$T = \$1216.65$

6. Use the laws of exponents to perform the following operations. ***(2 each)***

a. $x^2 * x^3$ x^5 b. $2^6 * 2^3$ 2^9

c. $3x * 5x^4$ $15x^5$ d. $6a^2b * -2ab^2$ $-12a^3b^3$

e. $\frac{x^5}{x^2}$ x^3 f. $\frac{11^7}{11^3}$ 11^4

g. $\frac{12x^6}{3x^2}$ $4x^4$ h. $\frac{-25x^4}{-5x^4}$ 5

i. $\frac{-27a^2b^4}{9a^3b^3}$ $\frac{-3b}{a}$ j. $(x^2)^5$ x^{10}

k. $(3x^5)^2$ $9x^{10}$ l. -5^2 -25

m. $(-2x^2y)^3$ $-8x^6y^3$

7. a. Evaluate each of the following. ***(2 each)***

i. $5 * 3^2 = 45$ ii. $4 + 6 * 3^2 = 58$

iii. $3 * 2^3 - 2 * 3^2 = 6$ iv. $5 + 2 * (7 - 3)^2 = 37$

b. Explain how exponentiation fits into the rules for order of operation. ***(6)***

The order of operation rules are as follows: Parentheses first, followed by exponentiation, followed by multiplication and division left to right, followed by addition and subtraction from left to right.

MATH *Connections* **I**
WELCOME TO ALGEBRA TEST (A)
CHAPTER 2

Name ______________________________ Date__________________

1. In 1993, the University of North Carolina defeated the University of Michigan in the finals of the NCAA men's basketball tournament, 77-71. The game was televised by NBC and had a TV household rating of 28.9%.

   a. There are three ways of scoring points in college basketball. A team could score a basket within 20 feet for two points, a basket further than 20 feet for three points or a foul shot for one point. Assign a variable to each scoring method. Explain why you chose that variable.

   b. Write an equation for determining the final score for a team if you know the number of two-point baskets, three-point baskets and foul shots.

   ____

   c. If you know that North Carolina scored 25 two-point baskets and 12 foul shots, write an equation for North Carolina's final score which will enable you to find the number of three-point baskets they scored.

   ____

   Solve the equation.

   d. In the equation, $71 = 3g + 50$

   g represents the number of three-point baskets Michigan scored. Solve the equation for g.

e. The total number of TV households in 1993, was 93,200,000 and 28.9% watched the game. How many TV households watched the game? Write your answer in scientific notation.

2. The freshmen class at Park School wants to have a class picnic in June. In September the class officers discussed fund raisers. They determine they need $25 per member of the class to cover all the expenses.
Li suggests a candy bar sale. He has found that they can buy ChocoMarvels for 68¢ and sell them for $1.25. They will need to advertise, distribute flyers and have other incidental costs. They estimate these to be $400.
Jane, the class treasurer, wonders how many ChocoMarvels each student will need to sell. Will each student have to sell too many bars? Should they consider other fund raisers? The questions below will help answer Jane's concerns.

   a. How much will it cost to send all 345 members of the freshmen class to the picnic? ________________

   b. How much will they need to raise to send the entire class and cover the advertising and other costs? ___________

   c. How much money will they make from the sale of each candy bar? _____

   d. Select a variable to represent the number of candy bars they need to sell.

   ____

   e. Jane wrote an expression to represent the money they will make from the sale of the ChocoMarvels. It is:

   ____

   f. Jane wrote an equation which, when solved, will tell them how many candy bars they must sell. The equation is:

   ____

   g. Solve the equation:

h. Estimate how many candy bars each member of the class must sell and explain how you determined your estimate.

____

____

____

i. Will each student have to sell too many bars? Should they consider other fund raisers? Jane has to write a report to the Executive Board concerning the ChocoMarvel fund raiser. Use the answers to the above questions to write what you feel Jane should say in the report.

____

____

____

____

____

3. Determine if the following are true or false:

_____ a. $(3.4) * (8.5) = (8.5) * (3.4)$

_____ b. $7 - 3 = 3 - 7$

_____ c. $3a + 12 = 12 + 3a$

_____ d. $3^2 = 2 * 2 * 2$

_____ e. $(1.04)^3 = (1.04) * (1.04) * (1.04) * (1.04)$

_____ f. $(a + b)^2 = a^2 + 2ab + b^2$

_____ g. $3(a + 6) = 3a + 6$

_____ h. $4^2 + 2^2 = (4 + 2)^2$

_____ i. 4.3 E7 = 430000000

_____ j. 0.00006 = 6E – 5

_____ k. 8200000000000 = 8.2E12

_____ l. $2 * 10^{-4} = 0.0002$

_____ m. If a = 2, then 5a – 6 = 46

_____ n. 5 + 3 * 8 = 29

_____ o. $3 * 7 + (8 - 5) \div 3 - 2 * 3^2 = 4$

4. Is the given number a solution of the equation? (Answer yes or no.)

   a. Is 3 a solution of 12 = 4a? __________

   b. Is 6 a solution of 15 = 3a + 5? __________

   c. Is 1 a solution of 8 = 5a + 8? __________

   d. Is 10 a solution of 25 = 5 + 2a? __________

Write an algorithm for determining whether a number is a solution for a given equation.

____

____

____

____

5. Use the laws of exponents to perform the following operations.

   a. $3x^4 * 2x^2$

   b. $6^4 * 6^2$

   c. 3^0

   d. $\frac{24x^5}{3x^2}$

   e. $\frac{70x^2}{10x^5}$

   f. $(4x^2)^3$

   g. $7x^0$

SOLUTION KEY AND SCORING GUIDE

MATH *Connections* **I**
WELCOME TO ALGEBRA TEST (A)
CHAPTER 2

1. In 1993, the University of North Carolina defeated the University of Michigan in the finals of the NCAA men's basketball tournament, 77-71. The game was televised by NBC and had a TV household rating of 28.9%.

a. There are three ways of scoring points in college basketball. A team could score a basket within 20 feet for two points, a basket further than 20 feet for three points or a foul shot for one point. Assign a variable to each scoring method. Explain why you chose that variable. ***(3)***

Answers will vary. Any choice would be correct.

b. Write an equation for determining the final score for a team if you know the number of two-point baskets, three-point baskets and foul shots. ***(3)***

Final Score = 2B + 3f + 1x *Students can use other variables but the constants have to be the same.*

c. If you know that North Carolina scored 25 two-point baskets and 12 foul shots, write an equation for North Carolina's final score which will enable you to find the number of three-point baskets they scored. ***(3)***

$77 = 25 * 2 + 12 * 1 + 3f$

Solve the equation. ***(3)***

$77 = 62 + 3f$
$15 = 3f$
$5 = f$

d. In the equation,

$$71 = 3g + 50$$

g represents the number of three-point baskets Michigan scored. Solve the equation for g. ***(3)***

$21 = 3g$
$7 = g$

f. The total TV households in 1993, was 93,200,000 and 28.9% of them watched the game. How many TV households watched the game? Write your answer in scientific notation. ***(3)***

$TV = .289 * 93\,200\,000$
$TV = 26\,934\,800$
$TV = 2.69348\text{E}7$

2. The freshmen class at Park School wants to have a class picnic in June. In September the class officers discussed fund raisers. They determine they need $25 per member of the class to cover all the expenses.
   Li suggests a candy bar sale. He has found that they can buy ChocoMarvels for 68¢ and sell them for $1.25. They will need to advertise, distribute flyers and have other incidental costs. They estimate these to be $400.
   Jane, class treasurer, wonders how many ChocoMarvels each student will need to sell. Will each student have to sell too many bars? Should they consider other fund raisers? The questions below will help answer Jane's concerns.

   a. How much will it cost to send all 345 members of the freshmen class to the picnic? _*345 * $25 = $8625*_ ***(3)***

   b. How much will they need to raise to send the entire class and cover the advertising and incidental costs? _*$9025*_ ***(3)***

   c. How much money will they make from the sale of each candy bar? _*57¢*_ ***(3)***

   d. Select a variable to represent the number of candy bars they need to sell. _*answers will vary*_ ***(2)***

   e. Jane wrote an expression to represent the money they will make from the sale of the ChocoMarvels. It is: ***(3)***
   _*Ex: .57c –$400*_

   f. Jane wrote an equation which, when solved, will tell them how many candy bars they must sell. The equation is: _*.57 c = 9025*_ ***(3)***

   g. Solve the equation: ***(3)***

   .57 c = 9025
   c = 15833.3 or 15834

   h. Estimate how many candy bars each member of the class must sell. Explain how you determined your estimate. ***(4)***

   Answers will vary. The estimates should be between 40 and 50. The most conventional way to estimate would be to divide 16000 by 400 and get 40. Since 400 is more than 345, the estimate should be between 40 and 50. The actual answer is approximately 46 candy bars.

i. Will each student have to sell too many bars? Should they consider other fund raisers? Jane has to write a report to the Executive Board concerning the ChocoMarvel fund raiser. Use the answers to the above questions to write what you feel Jane should say in the report. ***(6)***

A rich answer would contain an explanation about how realistic it is to have each class member sell between 40 and 50 candy bars. The expectation would be a definitive statement and reasons for the point of view.

3. Determine if the following are true or false: ***(2 each)***

T a. $(3.4) * (8.5) = (8.5) * (3.4)$

F b. $7 - 3 = 3 - 7$

T c. $3a + 12 = 12 + 3a$

F d. $3^2 = 2 * 2 * 2$

F e. $(1.04)^3 = (1.04) * (1.04) * (1.04) * (1.04)$

T f. $(a + b)^2 = a^2 + 2ab + b^2$

F g. $3(a + 6) = 3a + 6$

F h. $4^2 + 2^2 = (4 + 2)^2$

F i. 4.3 E7 = 430000000

T j. 0.00006 = 6E^{-}5

T k. 8200000000000 = 8.2E12

T l. $2 * 10^{-4} = 0.0002$

F m. If $a = 2$, then $5a - 6 = 46$

T n. $5 + 3 * 8 = 29$

T o. $3 * 7 + (8 - 5) \div 3 - 2 * 3^2 = 4$

4. Is the given number a solution of the equation? (Answer yes or no.) ***(2 each)***
   a. Is 3 a solution of $12 = 4a$? _*yes*_
   b. Is 6 a solution of $15 = 3a + 5$? _*no*_
   c. Is 1 a solution of $8 = 5a + 8$? _*no*_
   d. Is 10 a solution of $25 = 5 + 2a$? _*yes*_

   Write an algorithm for determining whether a number is a solution for a given equation.***(6)***

   A rich response should contain the idea that the possible solution is replaced for the variable in the equation. Then the necessary operations are carried out to determine if a true sentence occurs.

5. Use the laws of exponents to perform the following operations. ***(2 each)***

a.	$3x^4 * 2x^2$	$= \underline{6x^6}$	b.	$6^4 * 6^2$	$= \underline{6^6}$
c.	3^0	$= \underline{1}$	d.	$\frac{24x^5}{3x^2}$	$= \underline{8x^3}$
e.	$\frac{70x^2}{10x^5}$	$= \frac{7}{x^3}$	f.	$(4x^2)^3$	$= \underline{64x^6}$
g.	$(4x^2)^3$	$= \underline{7}$			

MATH *Connections* **I**
THE ALGEBRA OF STRAIGHT LINES QUIZ
SECTIONS 3.1 - 3.2 (A)

Name ______________________________________ Date ___________________

1. Zorba Stavapolous owns Mystic Pizza parlor in Gotham City. He delivers pizza in a region around his pizza parlor. In order to communicate quickly with his delivery people, Zorba decides to make a coordinate grid to cover the delivery region. Bus stops (B) are located at (4, 4) and (4, -4).

   - Mystic Pizza is located at the origin. Place a dot around the origin to show where Mystic Pizza is located.

   - The delivery region is bounded by the horizontal and vertical lines which connect the following four points: (12, 12), (12, -12), (-12, 12), (-12, -12). Put an X on each point and connect them with horizontal and vertical lines. Describe the region:

   ____

   ____

   ____

   Robin and Cardinal are the delivery persons for Mystic Pizza. Robin delivers north of the x-axis while Cardinal delivers south of the x-axis.

   Robin delivers pizzas to Penguin's home, then to Cat Woman's home. Name the coordinates of Penguin's home. ________ Name the coordinates of Cat Woman's home. __________

   Cardinal delivers pizzas to Joker's home, then to Anti-Math Man's home. Name the coordinates of Joker's home. _________ Name the coordinates of Anti-Math Man's home. _____________

   Cardinal calls Zorba from the payphone at (2,-7). Place a P at this point. Zorba wants her to pick up three pizzas and deliver them to a party at (-5, - 4). Place a C at this point. If the origin is moved to C, name the new coordinates of the point at which Penguin's home is located? ________

Robin calls Zorba from the payphone located at (-3,7). Place a T at this point. Zorba wants him to pick up one pizza and deliver it to Math Hero's home at (7, 10). Place an M at this point. If the origin is moved to M, name the new coordinates of the point at which Anti-Math's home is located?

____

If you delivered pizza for Mystic Pizza, explain which route you would prefer, Robin's or Cardinal's?

____

____

____

____

2. Write the set-builder notation for each of the line segments you drew to make the delivery region.

____

____

____

____

3. On the Gotham City map #2, draw the sets of points which are described as follows:

   a. the set of (x,y) such that $y = -5$ and x is between 0 and 5 inclusive.

   b. The set of (x,y) such that $y = 2$ and $x \geq -4$

   c. The set of (x,y) such that $x = 4$ and $-3 \leq y \leq 2$

   d. $\{ (x,6) \mid -1 \leq x \leq 5 \}$

   e. $\{ (-3, y) \mid 3 \leq y \leq 7 \}$

GOTHAM CITY #1

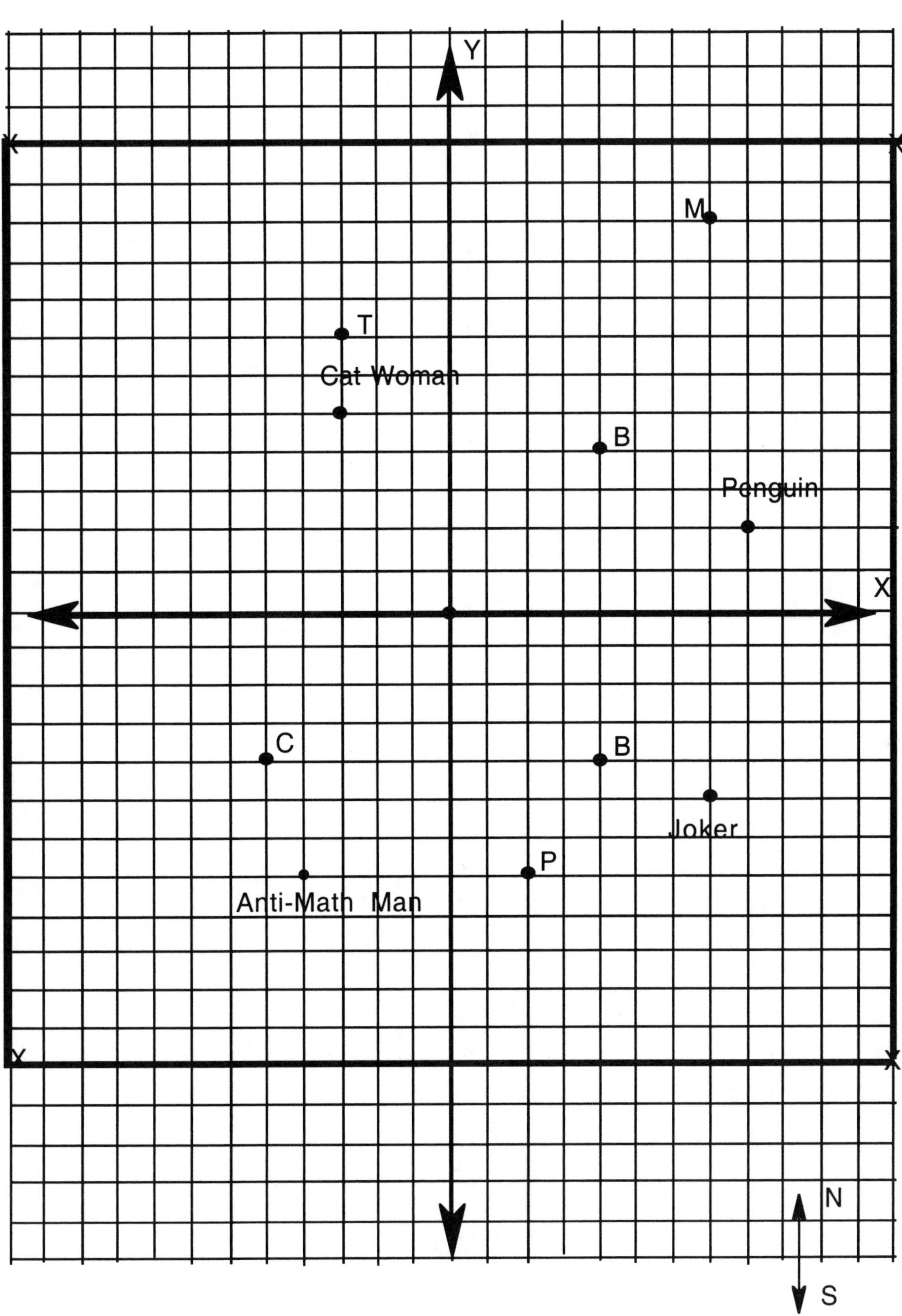

GOTHAM CITY #2

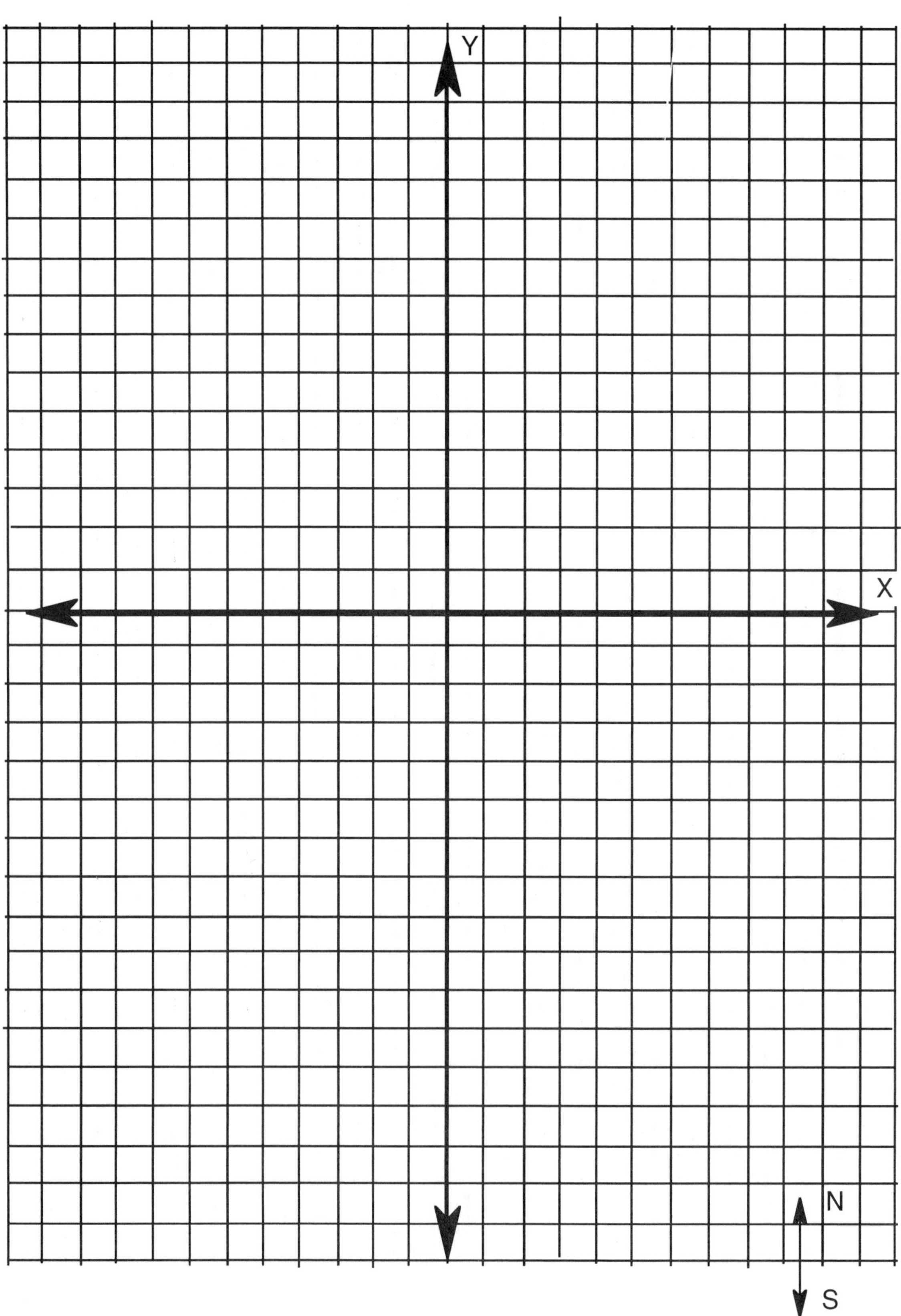

SOLUTION KEY AND SCORING GUIDE

MATH *Connections* **I**
THE ALGEBRA OF STRAIGHT LINES QUIZ
SECTIONS 3.1 - 3.2 (A)

1. Zorba Stavapolous owns Mystic Pizza parlor in Gotham City. He delivers pizza in a region around his pizza parlor. In order to communicate easily with his delivery persons Zorba decides to make a coordinate grid to cover the delivery region. Bus stops (B) are located at (4, 4) and (4, -4).

   - Mystic Pizza is located at the origin. Place a circle around the origin to show where Mystic Pizza is located. ***(4)***

   - The delivery region is bounded by the horizontal and vertical lines which connect the following four points: (12, 12), (12, -12), (-12, 12), (-12, -12). Put an X on each point and connect them with horizontal and vertical lines. ***(8)***

   Describe the region: ***(6)***

   The region is a square. Students may write more than this but the square shape is necessary to get full credit.

   Robin and Cardinal are the delivery persons for Mystic Pizza. Robin delivers north of the x-axis while Cardinal delivers south of the x-axis.

   Robin delivers pizzas to Penguin's home then Cat Woman's home. Name the coordinates of Penguin's home. *(8,2)* ***(4)*** Name the coordinates of Cat Woman's home. *(-3,5)* ***(4)***

   Cardinal delivers pizzas to Joker's home then to Anti-Math Man's home. Name the coordinates of Joker's home. *7,-5)* ***(4)*** Name the coordinates of Anti-Math Man's home. *(-4,-7)* ***(4)***

   Cardinal calls Zorba from the payphone at (2,-7). Place a P at this point. ***(4)*** Zorba wants her to pick up three pizzas and deliver them to a party at (- 5, - 4). Place a C at this point. ***(4)*** If the origin is moved to C, name the new coordinates of the point at which Penguin's home is located. *(13, 6)* ***(4)***

Robin calls Zorba from the payphone located at (-3,7). Place a T at this point. ***(4)*** Zorba wants him to pick up one pizza and deliver it to Math Hero's home at (7, 10). Place an M at this point. ***(4)*** If the origin is moved to M, name the new coordinates for Anti-Math's home. *(-11, -17)* ***(4)***

If you delivered pizza for Mystic Pizza, explain which route you would prefer, Robin's or Cardinal's? ***(6)***

The responses will be entirely opinion. A well written response with some supportive statements for explanation would get full credit.

2. Write the set-builder notation for each of the line segments you drew to make the delivery region. ***(5 each)***

   $\{(12, y) \mid -12 \le y \le 12\}$

   $\{(x, 12) \mid -12 \le x \le 12$

   $\{(-12, y) \mid -12 \le y \le 12\}$

   $\{(x, -12) \mid -12 \le x \le 12$

3. On the Gotham City map #2, draw the sets of points which are described as follows: ***(4 each - see map)***

   a. the set of (x,y) such that $y = -5$ and x is between 0 and 5 inclusive.

   b. The set of (x,y) such that $y = 2$ and $x \ge -4$

   c. The set of (x,y) such that $x = 4$ and $-3 \le y \le 2$

   d. $\{(x,6) \mid -1 \le x \le 5\}$

   e. $\{(-3, y) \mid 3 \le y \le 7\}$

GOTHAM CITY #1

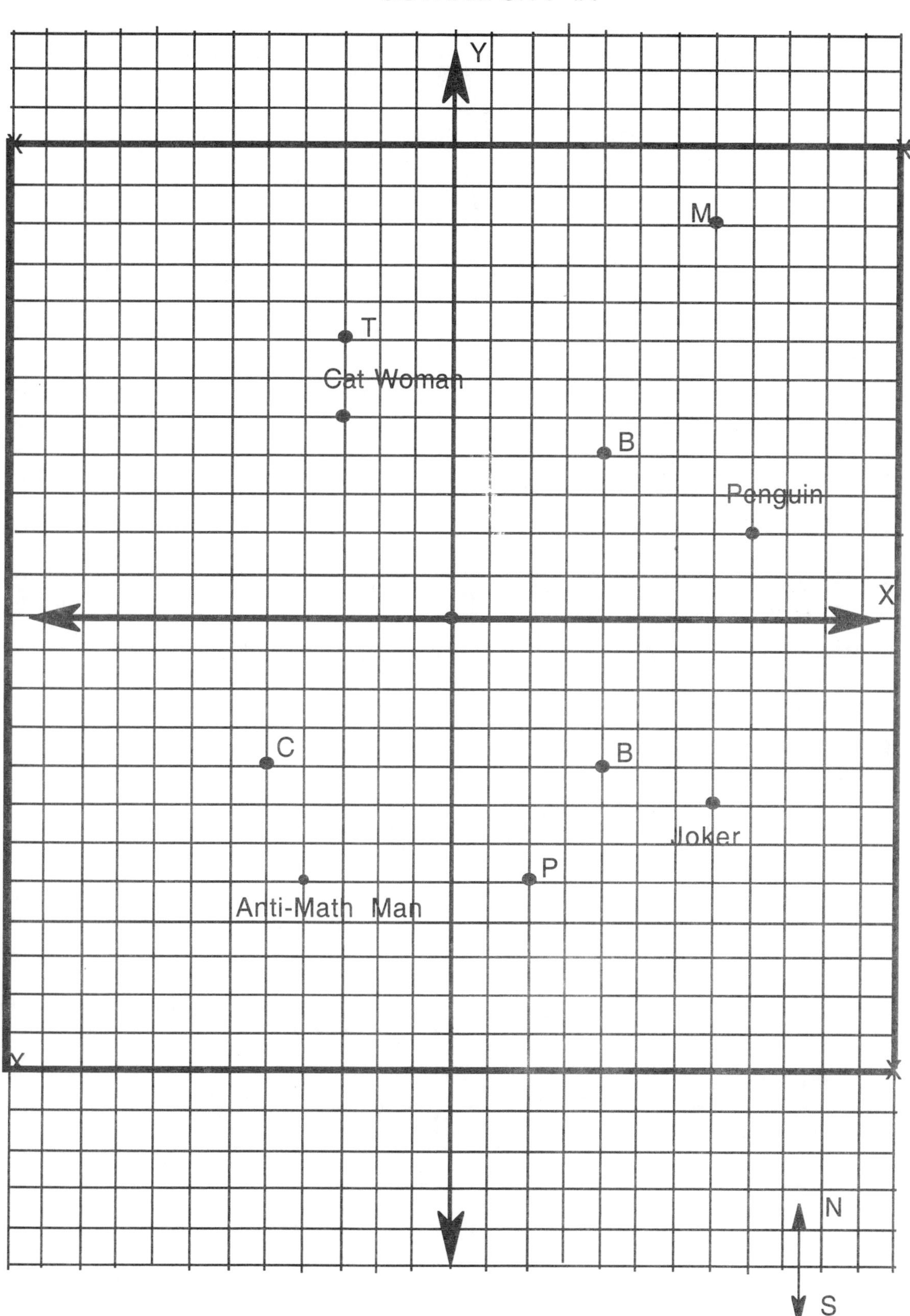

GOTHAM CITY #2

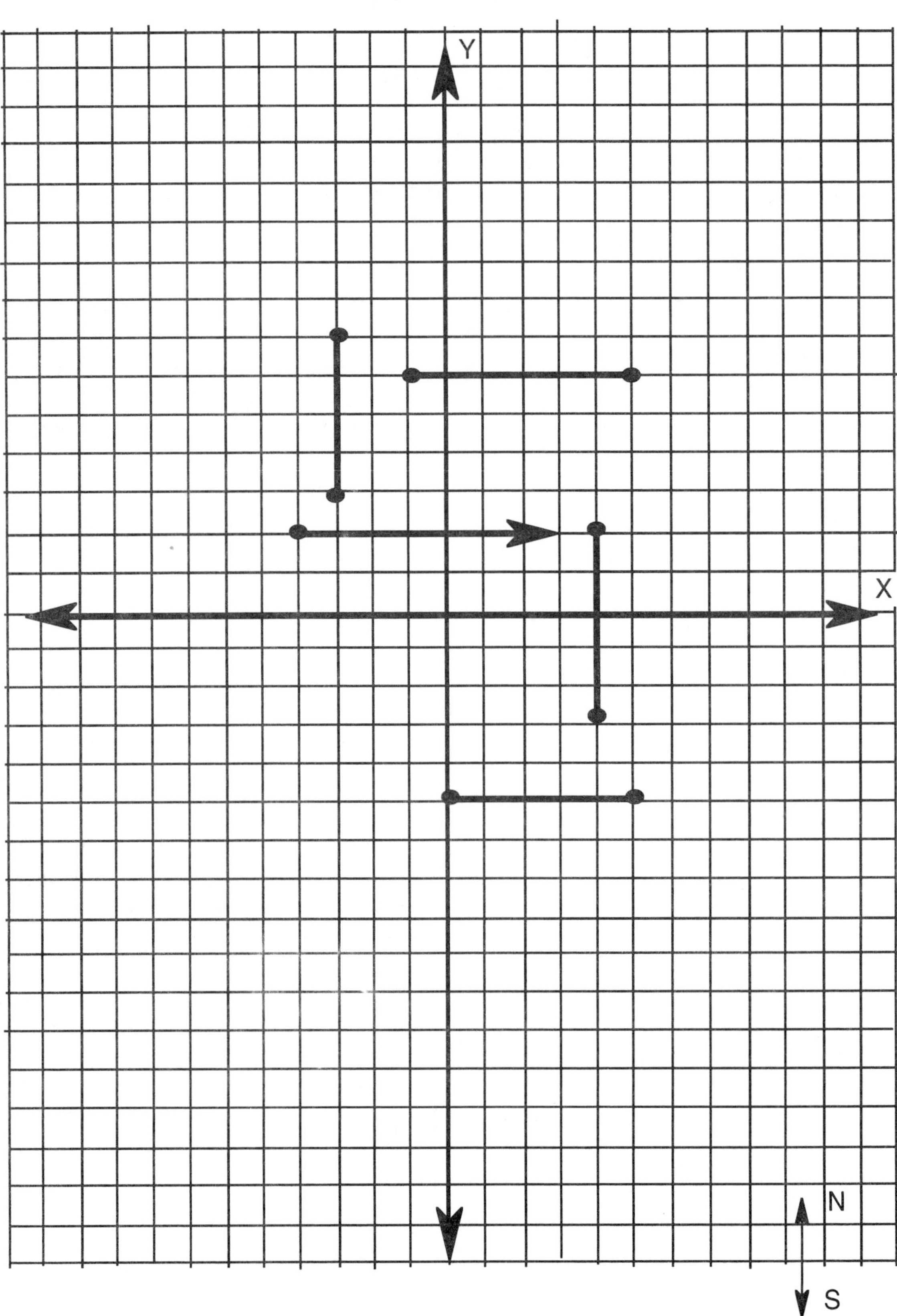

MATH *Connections* **I**
THE ALGEBRA OF STRAIGHT LINES QUIZ
SECTIONS 3.3-3.4 (A)

Name ________________________________ Date ______________

1. Zorba Stavapoulos owns Mystic Pizza parlor in Gotham City. He delivers pizza in a region around his pizza parlor. In order to communicate quickly with his delivery persons, Zorba decides to make a coordinate grid to cover the delivery region. Bus stops (B) are located at (4, 4) and (4, -4).

   - Mystic Pizza is located at the origin. Place a dot at the origin to show where Mystic Pizza is located.

   - The delivery region is bounded by the horizontal and vertical lines which connect the following four points: (12, 12), (12, -12), (-12, 12), (-12, -12). Put an X on each point and connect them with horizontal and vertical lines. Describe the region.

____

____

____

____

Mystic Pizza's main customers are Penguin, Joker, Cat Woman and Anti-Math Man. Name three coordinates at which each of their homes are located.

Penguin ______ Joker ______

Cat Woman ______ Anti-Math Man ______

To assist in the deliveries to their four homes, Mr. Stavapoulos wishes to develop an instant delivery system from the pizza parlor to their homes. In order to do this he will need the equations of the lines that connect the pizza parlor to their homes.

He asks Robin and Cardinal to determine the equation of each line. Robin and Cardinal recognize that they must finish the slope of the line from Mystic Pizza to each home before they can write the equation in $y = mx$ form. They organize their work below:

Line	Slope	Equation
Mystic to Penguin	__________	______________
Mystic to Joker	__________	______________
Mystic to Cat Woman	__________	______________
Mystic to Anti-Math Man	__________	______________

Complete Robin's and Cardinal's work in the box above.

The Futuristic Telecommunications Company tells Zorba that the delivery system will only work on the grid enclosed by the lines connecting (12, 0), (0, -12), (-12, 0), (0, 12). Mark these points on your Gotham City graph with T's and connect them to form a four-sided figure. Cardinal determines that the line connecting (12, 0) to (0, -12) has the equation y = 1x - 12. She enters the equation in the "Y = " menu on a graphing calculator and looks at the graph. Do the same on your calculator.

Cardinal determines the equations of the three other lines. Use a graphing calculator to find equations of the other three lines and write them in the blanks below:

Line connecting (0, 12) to (12, 0) ________________________

Line connecting (0, 12) to (-12, 0) ________________________

Line connecting (0, -12) to (12, 0) ________________________

2. You are given two equations below. Answer the following questions for these equations: y = 3x y = - 2x

- Give the coordinates of three points which are on the lines determined by each equation

$y = 3x$ $y = -2x$

_________ _________

_________ _________

_________ _________

- Give the coordinates of three points which are not on the lines determined by each equation.

$y = 3x$ $y = -2x$

_________ _________

_________ _________

_________ _________

- Which line has the steeper slope? __________ Explain your answer.

____

____

____

____

3. a. If (4, -1) are the coordinates of a point on a horizontal line, then what is the y-value of (5, y) if it is a point on the horizontal line? _________

b. What can be said about the coordinates of the points on a horizontal line?

____

____

____

____

c. What is the slope of a horizontal line? Explain how you determined your answer.

____

____

GOTHAM CITY

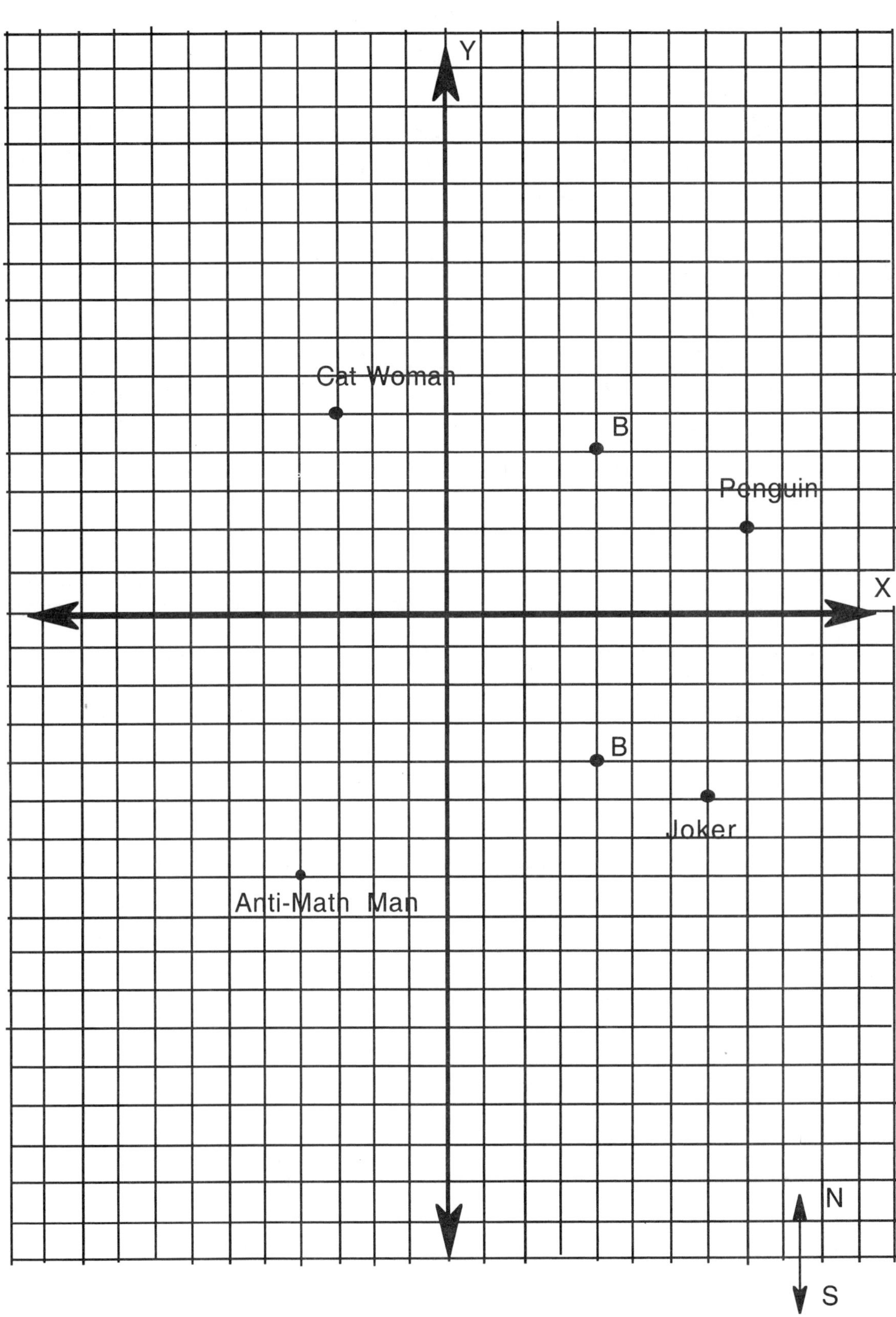

SOLUTION KEY AND SCORING GUIDE
MATH *Connections* **I**
THE ALGEBRA OF STRAIGHT LINES QUIZ
SECTIONS 3.3-3.4 (A)

1. Zorba Stavapoulos owns Mystic Pizza parlor in Gotham City. He delivers pizza in a region around his pizza parlor. In order to communicate quickly with his delivery persons, Zorba decides to make a coordinate grid to cover the delivery region. Bus stops (B) are located at (4, 4) and (4, -4).

   - Mystic Pizza is located at the origin. Place a dot at the origin to show where Mystic Pizza is located. ***(1)***

   - The delivery region is bounded by the horizontal and vertical lines which connect the following four points: (12, 12), (12, -12), (-12, 12), (-12, -12). Put an X on each point and connect them with horizontal and vertical lines. Describe the region. ***(3)***

     The region is a square with length of each side equal to 12.

   Mystic Pizza's main customers are Penguin, Joker, Cat Woman and Anti-Math Man. Name three coordinates at which each of their homes are located. ***(3 each)***

   Penguin *(8, 2)* Joker *(7, -5)*

   Cat Woman *(-3, 5)* Anti-Math Man *(-4, -7)*

To assist in the deliveries to their four homes, Mr. Stavapoulos wishes to develop an instant delivery system from the pizza parlor to their homes. In order to do this he will need the equations of the lines that connect the pizza parlor to their homes. He asks Robin and Cardinal to determine the equation of each line. Robin and Cardinal recognize that they must finish the slope of the line from Mystic Pizza to each home before they can write the equation in $y = mx$ form. They organize their work below: ***(6 each or 24 total)***

Line	Slope	Equation
Mystic to Penguin	$\frac{2}{8}$ or $\frac{1}{4}$	$y = \frac{1}{4}x$
Mystic to Joker	$-\frac{5}{7}$	$y = -\frac{5}{7}x$
Mystic to Cat Woman	$-\frac{5}{3}$	$y = -\frac{5}{3}x$
Mystic to Anti-Math Man	$\frac{7}{4}$	$y = \frac{7}{4}x$

Complete Robin's and Cardinal's work in the box above.

The Futuristic Telecommunications Company tells Zorba that the delivery system will only work on the grid enclosed by the lines connecting (12, 0), (0, -12), (-12, 0), (0, 12). Mark these points on your Gotham City graph with T's and connect them to form a four-sided figure. Cardinal determines that the line connecting (12, 0) to (0, -12) has the equation $y = 1x - 12$. She enters the equation in the "Y = " menu on a graphing calculator and looks at the graph. Do the same on your calculator.

Cardinal determines the equations of the three other lines. Use a graphing calculator to find equations of the other three lines and write them in the blanks below: ***(6 each)***

Line connecting (0, 12) to (12, 0) $y = -1x + 12$

Line connecting (0, 12) to (-12, 0) $y = 1x + 12$

Line connecting (0, -12) to (12, 0) $y = -1x - 12$

2. You are given two equations below. Answer the following questions for these equations: y = 3x y = - 2x

- Give the coordinates of three points which are on the lines determined by each equation. ***(6)***

  y = 3x y = - 2x

  ________ ________

  Answers will vary.

  ________ ________

- Give the coordinates of three points which are not on the lines determined by each equation. ***(6)***

  y = 3x y = - 2x

  ________ ________

  Answers will vary.

  ________ ________

- Which line has the steeper slope? *y = 3x* ***(4)*** Explain your answer. ***(6)***

The line that has a slope in which the horizontal rise is more than the vertical run has the steeper slope. In this case 3:1 is greater than -2:1.

3. a. If (4, -1) are the coordinates of a point on a horizontal line, then what is the y- value of (5, y) if it is a point on the horizontal line? *- 1* ***(4)***

   b. What can be said about the coordinates of the points on a horizontal line? ***(6)***

   All the y coordinates are the same on a horizontal line.

   c. What is the slope of a horizontal line? *0* Explain how you determined your answer. ***(6)***

   Since the change in y on any horizontal line is always 0, the slope is 0 because zero divided by any number is zero.

GOTHAM CITY

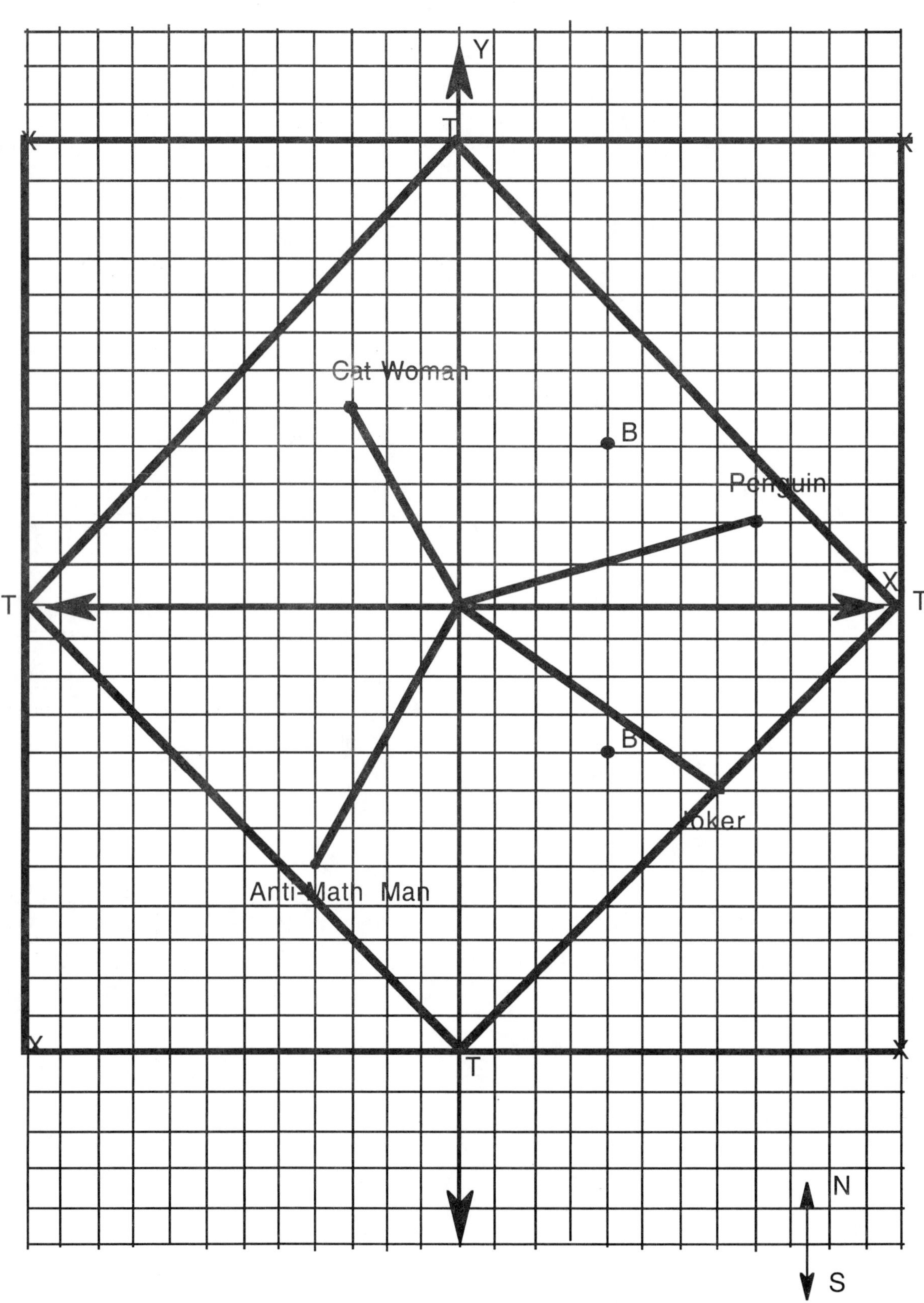

MATH *Connections* I
THE ALGEBRA OF STRAIGHT LINES QUIZ
SECTIONS 3.5-3.6 (A)

Name ____ Date ________________

1. Zorba Stavapolous owns Mystic Pizza parlor in Gotham City. He delivers pizza in a region around his pizza parlor. In order to communicate quickly with his delivery people Zorba decides to make a coordinate grid to cover the delivery region. Bus stops (B) are located at (4, 4) and (4, -4).

   - Mystic Pizza is located at the origin. Place a circle around the origin to show where Mystic Pizza is located.
   - The delivery region is bounded by the horizontal and vertical lines which connect the following four points: (12, 12), (12, -12), (-12, 12), (-12, -12). Put an X on each point and connect them with horizontal and vertical lines. Describe the region.

   ____

   ____

   ____

   Mystic Pizza's main customers are Penguin, Joker, Cat Woman, and Anti-Math Man. Name the coordinates of the points at which each of their homes are located.

   Penguin ______ Joker ______

   Cat Woman ______ Anti-Math Man ______

   When Penguin, Joker, Cat Woman and Anti-Math Man discover that the Futuristic Tele-Communications Company (FTC) has developed an instant communications system, they want to join in order to harass Batman. The FTC is only able to provide them with 3 links. It is decided to link Penguin to Joker, Joker to Cat Woman and Cat Woman to Anti-Math Man.

Joker is placed in charge of determining the equations that will provide the three links. In order to do this he needs to find the slope and y-intercept of the line that makes each link. He will then write each equation in $y = mx + b$ form. He organizes his work below:

Line Connecting	Slope	Y-intercept	Equation: y = mx + b
Penguin to Joker	_____	_______	_________________
Cat Woman to Joker	_____	_______	_________________
Penguin to Anti-Math Man	_____	_______	_________________

Joker decides to attempt to steal an additional line to connect him to Anti-Math Man. He determines the equation of the link is $y = 0.2x - 7$. Is he correct? _______ If Joker is correct show the work he used to obtain the answer. If Joker is not correct, show the work to determine the correct equation for the link.

2. a. Find the equation for the line which contains the points (1, -4) and (5, 2). Show all work.

   b. Find the equation of the line that contains the points (0, 1) and (-4, 5). Show all work.

   c. For which problem was it easier to determine the equation of the line? _______ Explain your answer.

   ____

   ____

   ____

3. The Manhattan Public Works Department has to build a bridge over a stream in the park. The width of the stream at the spot where the bridge is to be constructed is 60 feet. The bridge will rise 6 feet from one side of the stream to the other. Support poles will be placed every 10 feet. Your job is to cut the support poles the proper lengths. Determine the lengths. Make a diagram to assist you.

GOTHAM CITY

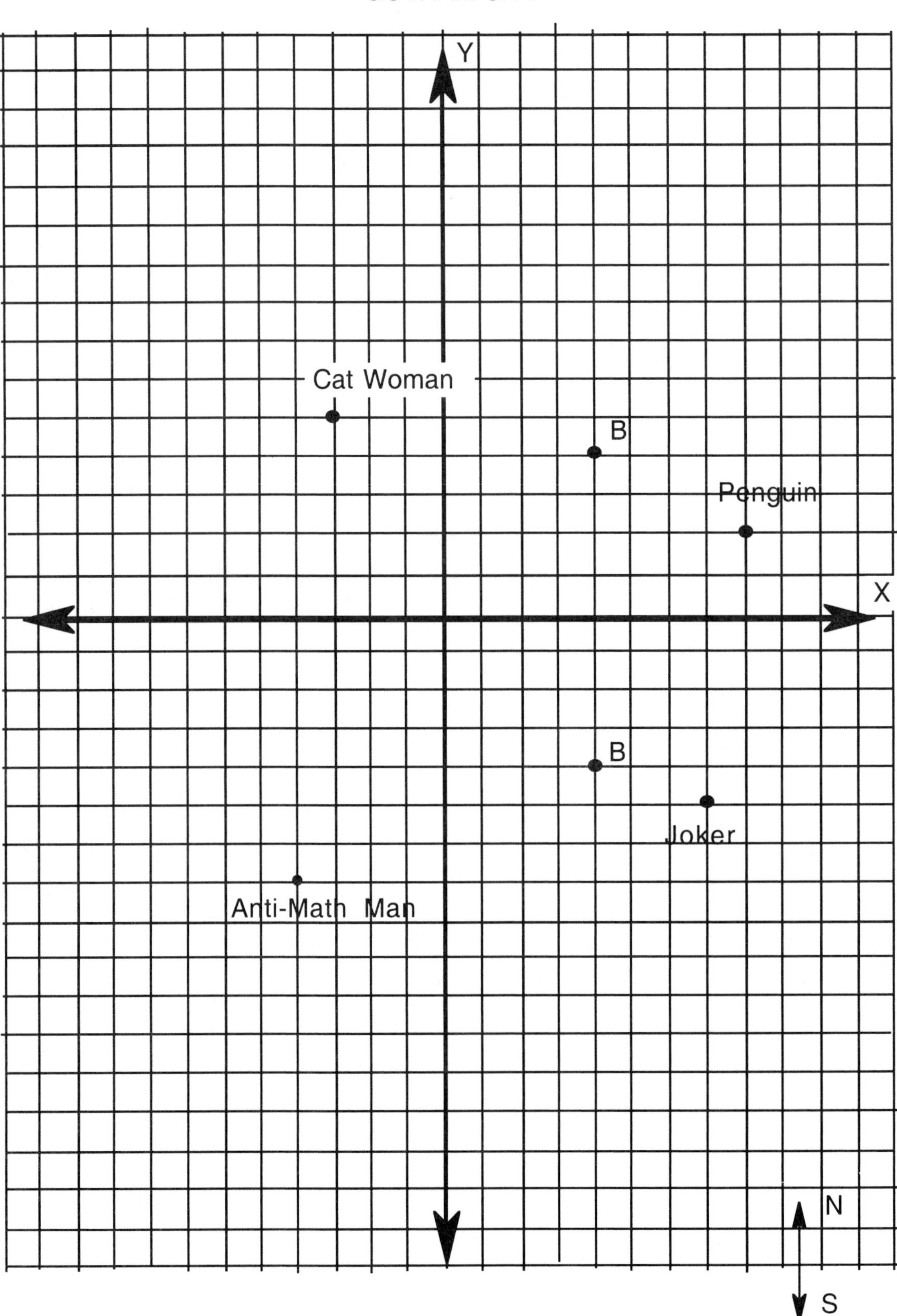

SOLUTION KEY AND SCORING GUIDE

MATH *Connections* **I**
THE ALGEBRA OF STRAIGHT LINES QUIZ
SECTIONS 3.5-3.6 (A)

1. Zorba Stavapolous owns Mystic Pizza parlor in Gotham City. He delivers pizza in a region around his pizza parlor. In order to communicate quickly with his delivery people Zorba decides to make a coordinate grid to cover the delivery region. Bus stops (B) are located at (4, 4) and (4, -4).

   - Mystic Pizza is located at the origin. Place a circle around the origin to show where Mystic Pizza is located.
   - The delivery region is bounded by the horizontal and vertical lines which connect the following four points: (12, 12), (12, -12), (-12, 12), (-12, -12). Put an X on each point and connect them with horizontal and vertical lines. Describe the region.

     The region is a square with length of each side equal to 12.

   Mystic Pizza's main customers are Penguin, Joker, Cat Woman and Anti-Math Man. Name three coordinates at which each of their homes are located. ***(1 each)***

   Penguin *(8, 2)* Joker *(7, -5)*

   Cat Woman *(-3, 5)* Anti-Math Man *(-4, -7)*

   When Penguin, Joker, Cat Woman and Anti-Math Man discover that the Futuristic Tele-Communications Company (FTC) has developed an instant communications system, they want to join in order to harass Batman. The FTC is only able to provide them with 3 links. It is decided to link Penguin to Joker, Joker to Cat Woman and Cat Woman to Anti-Math Man.

Joker is placed in charge of determining the equations that will provide the three links. In order to do this he needs to find the slope and y-intercept of the line that makes each link. He will then write each equation in $y = mx + b$ form. He organizes his work below:

Line Connecting	Slope	Y-intercept	Equation: y = mx + b	
Penguin to Joker	*7*	*- 54*	$y = 7x - 54$	**(15)**
Cat Woman to Joker	*- 1*	*2*	$y = -x + 2$	**(15)**
Penguin to Anti-Math Man	*12*	*41*	$y = 12x + 41$	**(15)**

Joker decides to attempt to steal an additional line to connect him to Anti-Math Man. He determines the equation of the link is y = 0.2x - 7. Is he correct? *no* ***(5)*** If Joker is correct show the work he used to obtain the answer. If Joker is not correct, show the work to determine the correct equation for the link. ***(10)***

Slope is $\frac{2}{11}$

Y-intercept is $\frac{69}{11}$

Equation is $y = \frac{2}{11}x \quad \frac{69}{11}$

2. a. Find the equation for the line which contains the points (1, -4) and (5, 2). Show all work. ***(10)***

$m = \frac{3}{2}$

y-intercept = $-\frac{11}{2}$

Equation is: $y = \frac{3}{2}x - \frac{11}{2}$

b. Find the equation of the line that contains the points (0, 1) and (-4, 5). Show all work. ***(10)***

$m = -1$

Equation is: $y = -1x + 1$

c. For which problem was it easier to determine the equation of the line? <u>*b*</u>
Explain your answer. ***(6)***

In problem b, one ordered pair is given in the form (0, B) so that the y-intercept is immediately known.

3. The Manhattan Public Works Department is to build a bridge over a stream in the park. The width of the stream at the spot where the bridge is to be constructed is 60 feet. The bridge will rise 6 feet from one side of the stream to the other. Support poles will be placed every 10 feet. Your job is to cut the support poles the proper lengths. Determine the lengths. Make a diagram to assist you. ***(10)***

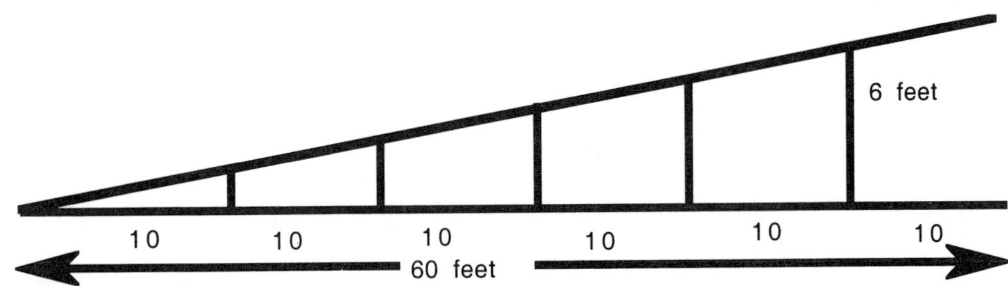

GOTHAM CITY

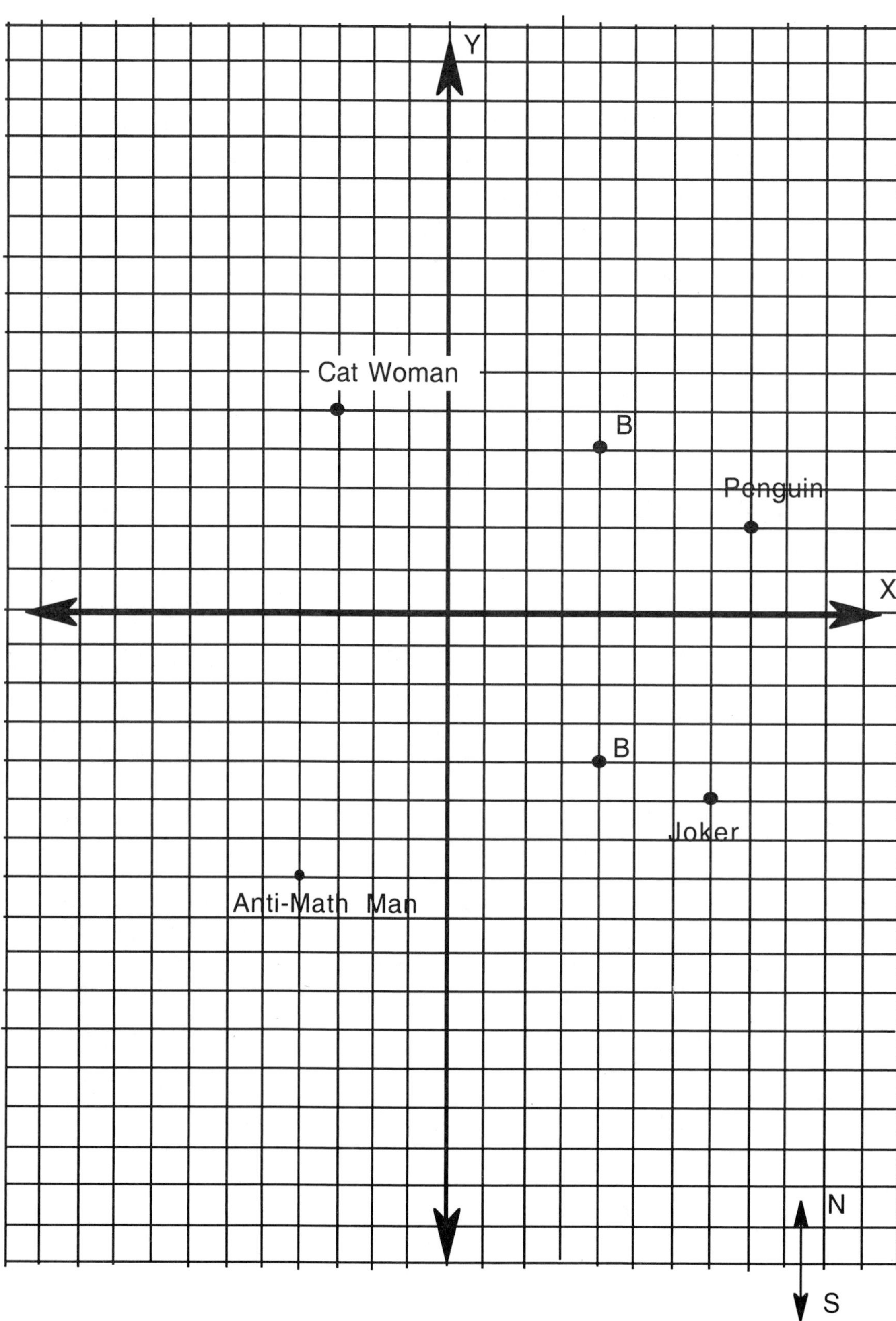

MATH *Connections* **I**
THE ALGEBRA OF STRAIGHT LINES TEST (A)
CHAPTER 3

Name ______________________________ Date ____________

1. Taquisha is in charge of organizing a scavenger hunt for the final activity for the summer playground. Since the young people on the playground will be going back to school soon, she decides to relate the scavenger hunt clues to a coordinate system so they will review some math before school starts. If you are able to follow all the clues you will find all the prizes and will be led to the grand prize which is a membership in the Ice Cream Sundae of the Month Club.
The corners of the pool are located at (0, 4), (0, -4), (3, -4) and (3, 4).

   a. Place X's on points (-2, 0) and (0, 5). Bags of M&M's are located at each point.

   b. Draw the line connecting (-2, 0) and (0,5).

   c. The slope of the line you drew is: __________ (Write the slope as a common fraction in lowest terms.)

   d. Use the denominator of the slope (change in x) as the x-coordinate and the numerator (change in y) as the y-coordinate to form an ordered pair.

   e. That ordered pair is: _______ Graph it on the coordinate system with a O around the point. Pizza coupons are located there.

   f. Place a ▭ on the points (-3, -3) and (1, 6). Reese's peanut butter cups are located at each point.

   g. Draw a line connecting (-3, -3) and (1, 6).

   h. The slope of the line you drew is: _______. (Write the slope as a common fraction in lowest terms.)

   i. Use the denominator of the slope (change in x) as the x-coordinate and the numerator (change in y) as the y-coordinate to form an ordered pair.

   j. That ordered pair is: _______ Graph it on the coordinate system with a O around the point. Movie rental coupons are located here.

k. Draw a line through the two (O) symbols. The equation of that line is:

____

l. Make an ordered pair whose x-coordinate is the slope of the line and whose y-coordinate is the y-intercept of the line. The ordered pair is______.

m. Taquisha found that if she placed the grand prize at this point it would be in the swimming pool. Therefore she changed the origin to (2, -8) and marked off the ordered pair in part l. from there. Place a G at that point. That is where the grand prize is hidden.

2. To use the instant delivery service from Mystic Pizza, Zorba Stavapolous has a delivery charge of $1.50 for each order he sends through the link. Each pizza costs $8.25. The instant delivery link can handle only one size. If someone ordered 3 pizzas the cost of the order would be determined as follows:
   3 (8.25) + 1.50 = 26.25.

   a. Write an equation or formula in the form of $y = mx + b$ for determining the cost of delivering any number of pizzas:

   ____

   b. On a calculator from the ZOOM function choose the ZINTEGER screen (ENTER) to graph your equation. Does the line go through the origin? ________ Explain.

   ____

   ____

   c. Use the TRACE function to find the cost of delivering the following number of pizzas:

   1 pizza ____________________

   5 pizzas ____________________

   8 pizzas ____________________

   d. In the equation you wrote for part a, what does the slope mean in terms of the cost of the pizzas?

   ____

   ____

In the equation you wrote for part a, what does the y-intercept mean in terms of the cost of the pizzas?

____

____

3.

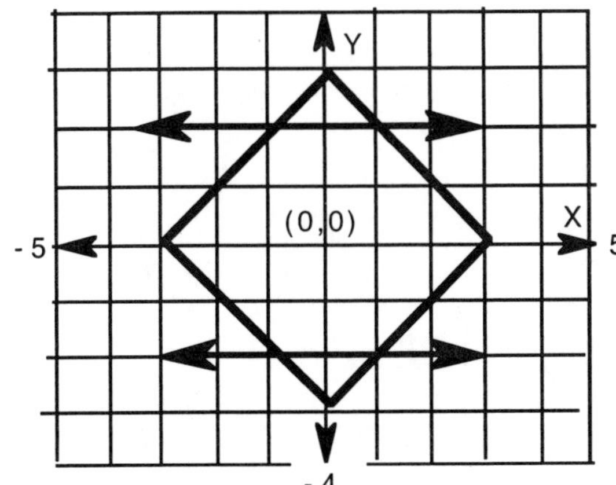

List the 6 equations that would be used to make the design above:

1.________________ 4. ______________

2.________________ 5. ______________

3.________________ 6. ______________

4. • Write an equation in the spaces below that satisfies the directions.
   • Graph each equation on a calculator.
   • Copy the graphs on to the graph paper provided by your teacher.

DIRECTIONS: Using only numbers between -10 and +10 write equations that satisfy the following requirements:

1. A line that contains the origin. ______________________________
2. A line with y-intercept of -2 and slope that is greater than 2. ____________
3. A line with a negative slope that does not contain the origin. ____________
4. A line parallel to the line in equation 1. ______________________

5. On the graph paper provided by your teacher, graph the sets of points that satisfy the following:

   a. {(x,y)| -2 x 5 and y = 3}

   b. {(x,y)| - 2 y 3 and x = - 4}

   c. {(x,y)| y 2 and x = 5 }

6. Identify the set of points that would describe the graphs below:

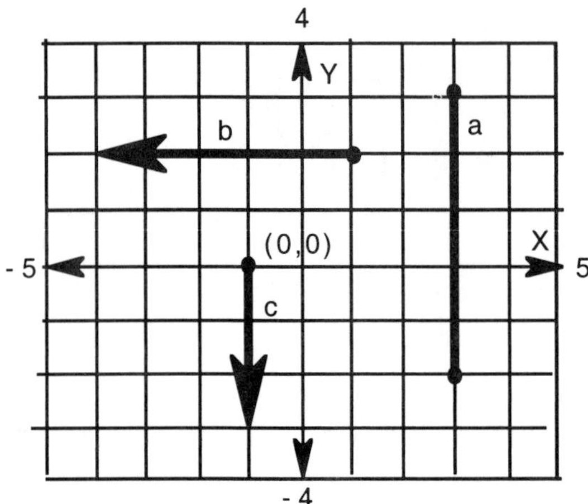

a. ______________________________

b. ______________________________

c. ______________________________

SCAVENGER HUNT

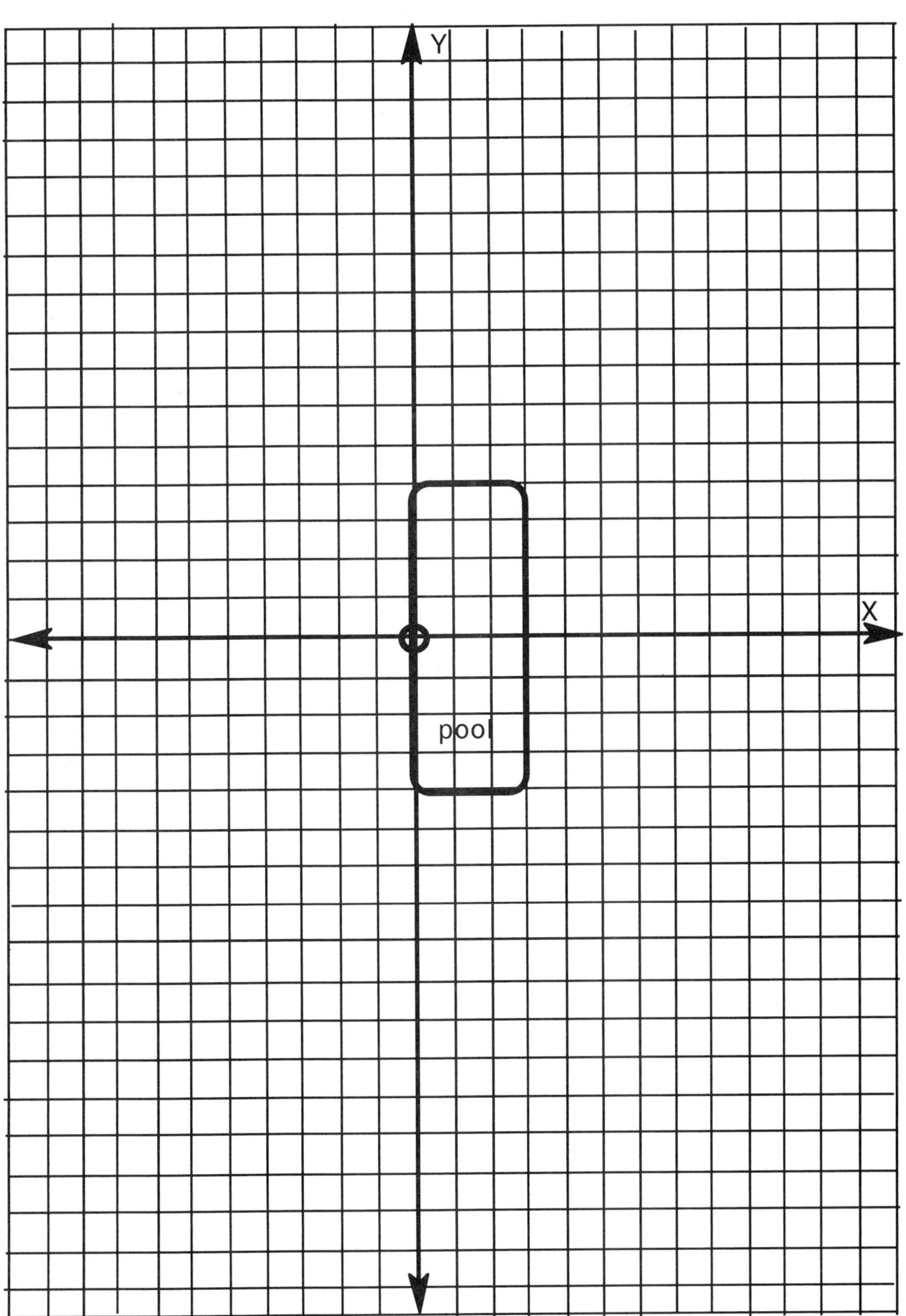

SOLUTION KEY AND SCORING GUIDE

MATH *Connections* **I**
THE ALGEBRA OF LINES TEST (A)
CHAPTER 3

1. Taquisha is in charge of organizing a scavenger hunt for the final activity for the summer playground. Since the young people on the playground will be going back to school soon, she decides to relate the scavenger hunt clues to a coordinate system so they will review some math before school starts. If you are able to follow all the clues you will find all the prizes and will be led to the grand prize which is a membership in the Ice Cream Sundae of the Month Club.

   The corners of the pool are located at (0, 4), (0, -4), (3, -4) and (3, 4).

   a. Place X's on points (-2, 0) and (0, 5). Bags of M&M's are located at each point. ***(2)***

   b. Draw the line connecting (-2, 0) and (0,5).

   c. The slope of the line you drew is: $\frac{5}{2}$ (Write the slope as a common fraction in lowest terms.) ***(2)***

   d. Use the denominator of the slope (change in x) as the x-coordinate and the numerator (change in y) as the y-coordinate to form an ordered pair.

   e. That ordered pair is: *(2, 5)* Graph it on the coordinate system with a ○ around the point. Pizza coupons are located there. ***(2)***

   f. Place a ▭ on the points (-3, -3) and (1, 6). Reese's peanut butter cups are located at each point. ***(2)***

   g. Draw a line connecting (-3, -3) and (1, 6).

   h. The slope of the line you drew is: $\frac{9}{4}$. (Write the slope as a common fraction in lowest terms.) ***(2)***

   i. Use the denominator of the slope (change in x) as the x-coordinate and the numerator (change in y) as the y-coordinate to form an ordered pair.

   j. That ordered pair is: *(4, 9)* Graph it on the coordinate system with a ○ around the point. Movie rental coupons are located here. ***(2)***

k. Draw a line through the two circles (O). The equation of that line is:

y = 2 x + 1 ***(3)***

l. Make an ordered pair whose x-coordinate is the slope of the line and ***(2)*** whose y-coordinate is the y-intercept of the line. The ordered pair is *(2, 1)* .

m. Taquisha found that if she placed the grand prize at this point it would be in the swimming pool. Therefore she changed the origin to (2, -8) and marked off the ordered pair in part l. from there. Place a G at that point. That is where the grand prize is hidden. ***(3)***

2. To use the instant delivery service from Mystic Pizza, Zorba Stavapolous has a delivery charge of \$1.50 for each order he sends through the link. Each pizza is \$8.25. The instant delivery link can handle only one size. If someone ordered 3 pizzas the cost of the order would be determined as follows: 3 (8.25) + 1.50 = 26.25.

a. Write an equation or formula in the form of $y = mx + b$ for determining the cost of delivering any number of pizzas: *y = 8.25 x + 1.50* ***(5)***

b. On a calculator from the ZOOM function choose the ZINTEGER screen (ENTER) to graph your equation. Does the line go through the origin? *no* ***(3)*** Explain.

The line does not go through the origin because the equation is in y = mx + b form instead of y = mx form. ***(6)***

c. Use the TRACE function to find the cost of delivering the following number of pizzas: ***(6)***

1 pizza *\$9.75*

5 pizzas *\$42.75*

8 pizzas *\$67.50*

d. In the equation you wrote, what does the slope mean in terms of the cost of the pizzas? ***(6)***

The slope is the price per pizza.

In the equation you wrote, what does the y-intercept mean in terms of the costs of the pizzas? ***(6)*** *The y-intercept represents the cost to deliver the pizzas.*

3.

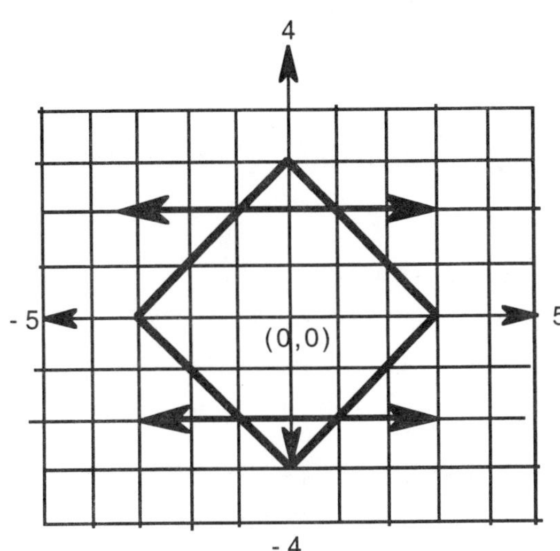

List the 6 equations that would be used to make the design above: ***(3 each)***

1. *$y = 2$*
2. *$y = x - 3$*
3. *$y = x + 3$*
4. *$y = -x - 3$*
5. *$y = -x + 3$*
6. *$y = -2$*

4. • Write an equation in the spaces below that satisfies the directions.
   • Graph each equation on a calculator.
   • Copy the graphs on to the graph paper provided by your teacher.

DIRECTIONS: Using only numbers between -10 and +10, write equations that satisfy the following requirements: ***(3 each)***

1. A line that contains the origin. *A line in the form of $y = mx$*
2. A line with y-intercept of -2 and slope that is greater than 2. *Ex: $y = 3x - 2$*
3. A line with a negative slope that does not contain the origin. *Ex: $y = -1x + 1$*
4. A line parallel to the line in equation 1. *A line in the form of $Y = mx + b$*

5. On the graph paper provided by your teacher, graph the sets of points that satisfy the following: ***(3each)***

   a. {(x,y)| -2   x   5 and y = 3}

   b. {(x,y)| - 2 y 3 and x = - 4}

   c. {(x,y)| y 2 and x = 5 }

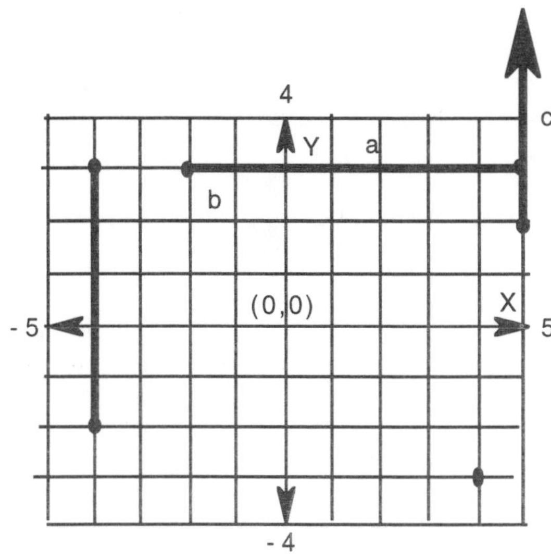

6. Identify the set of points that would describe the graphs below: ***(3 each)***

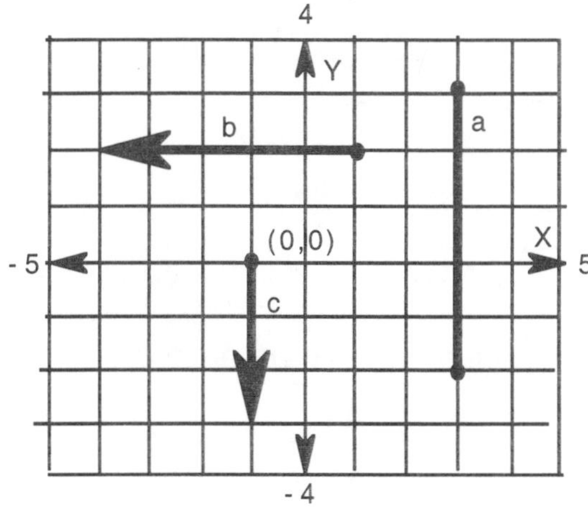

a. *{(3 , y) | - 2   y   3}* b. *{ (x , 2) | x 1}*

c. *{ (- 1, y) | y 0 }*

SCAVENGER HUNT

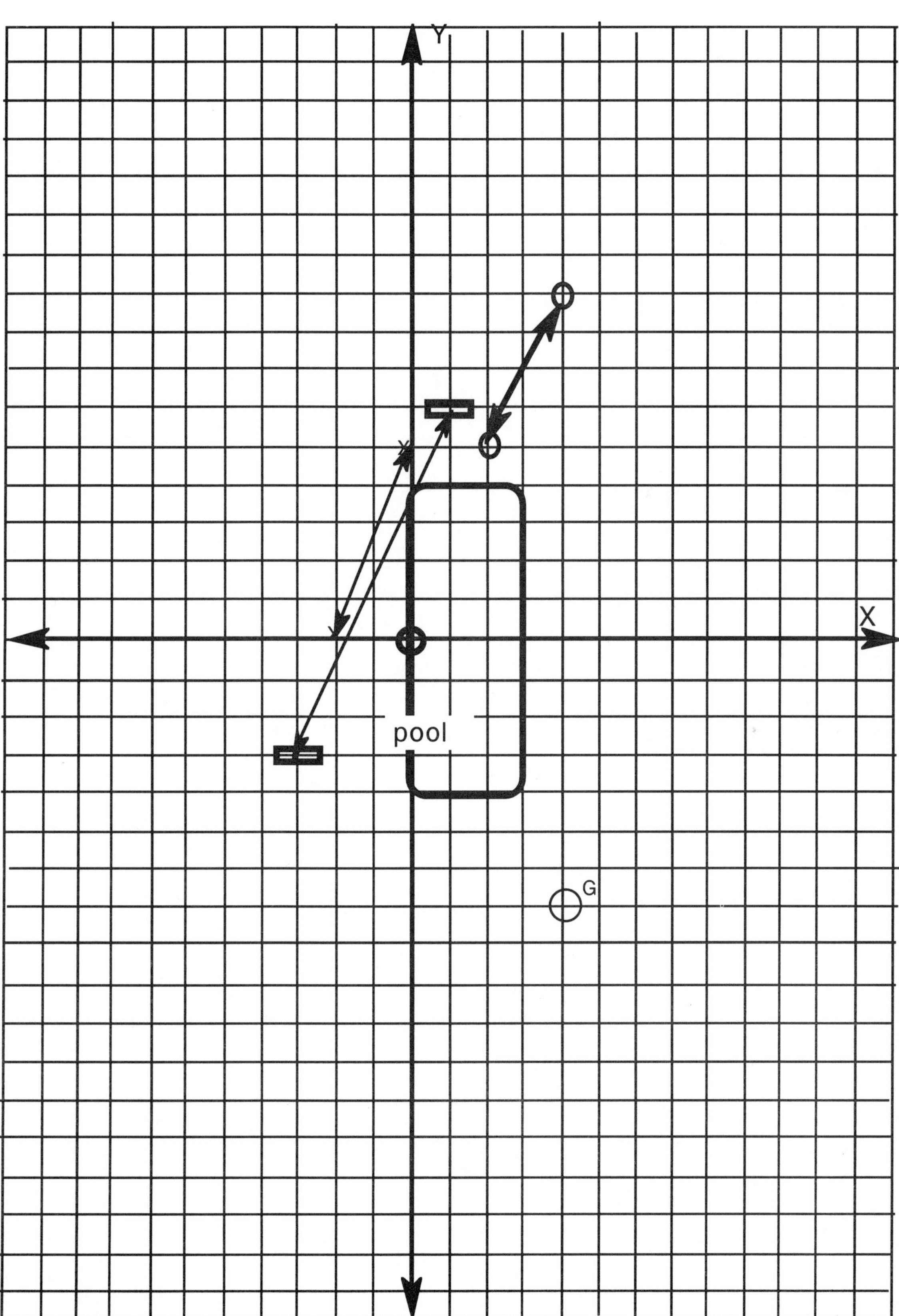

MATH *Connections* **I**
GRAPHICAL ESTIMATION QUIZ
SECTIONS 4.1 - 4.2 (A)

Name ______________________________ Date ____________

1. The number of basketball coaches that reside in the six states that contain the geographical area of Southern New England and the Northern Mid-Atlantic States is listed in the table below:

COACHES

STATE	NUMBER OF FEMALE COACHES	NUMBER OF MALE COACHES
CONNECTICUT	1298	2450
MASSACHUSETTS	3630	6715
NEW JERSEY	2784	5102
NEW YORK	7334	14665
PENNSYLVANIA	4600	12785
RHODE ISLAND	580	1105

a. Use the following directions:
- Make a scattergram for this data.
- Use the horizontal axis for the number of female coaches using a scale from 500 to 7,500.
- Use the vertical axis for the number of male coaches using a scale from 1000 to 15,000.

b. Is there a pattern displayed by the points on the scattergram? Explain your answer.

____

____

____

2. MADD (Mothers Against Drunk Driving) is a national organization that provides information and educational programs to assist local groups and governments to provide programs to make the public aware of the problems with drunk drivers. Melissa Hargrove, president of a local chapter of MADD, researched the fatalities from automobile accidents from 1966 to 1990. She presented the following data to her chapter.

TRAFFIC FATALITIES

YEAR	FATALITY RATES PER 100 000 DRIVERS
1966	50
1970	47
1974	36
1978	36
1982	29
1986	29
1990	27

The scattergram for this data is displayed below:

TRAFFIC FATALITIES / 100,000 DRIVERS

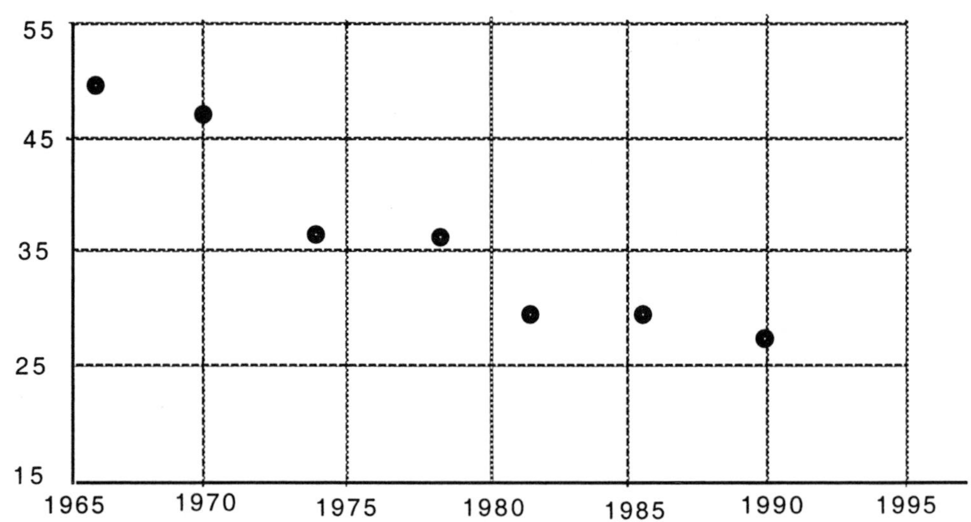

a. Lucy Winfield wondered what the fatality rate was in 1973. In order to find this from the graph, linear interpolation was used. Use a ruler to demonstrate how you can find an estimate for the number of fatalities in 1973 using linear interpolation. What estimate did you find? ____________________

b. When using linear interpolation, what does your answer represent?

____

____

____

c. Find the equation of the line that contains the line segment connecting (70, 47) to (74, 36). Show your work.

d. Use the equation from part c. to find an estimate for the number of fatalities in 1973 by substituting 73 into the equation. Show your work.

e. How did the your estimates from parts a. and d. compare? Be as specific as you can.

____

____

____

f. Write **at least two reasons** why you think fatalities decreased from 1966 to 1990.

____

____

____

____

SOLUTION KEY AND SCORING GUIDE

MATH *Connections* **I**
GRAPHICAL ESTIMATION QUIZ
SECTIONS 4.1 - 4.2 (A)

1. The number of basketball coaches that reside in the six states that contain the geographical area of Southern New England and the Northern Mid-Atlantic States is listed in the table below:

	COACHES	
STATE	NUMBER OF FEMALE COACHES	NUMBER OF MALE COACHES
CONNECTICUT	1298	2450
MASSACHUSETTS	3630	6715
NEW JERSEY	2784	5102
NEW YORK	7334	14665
PENNSYLVANIA	4600	12785
RHODE ISLAND	580	1105

a. Use the following directions: ***(25)***
   - Make a scattergram for this data.
   - Use the horizontal axis for the number of female coaches using a scale from 500 to 7,500.
   - Use the vertical axis for the number of male coaches using a scale from 1000 to 15,000.

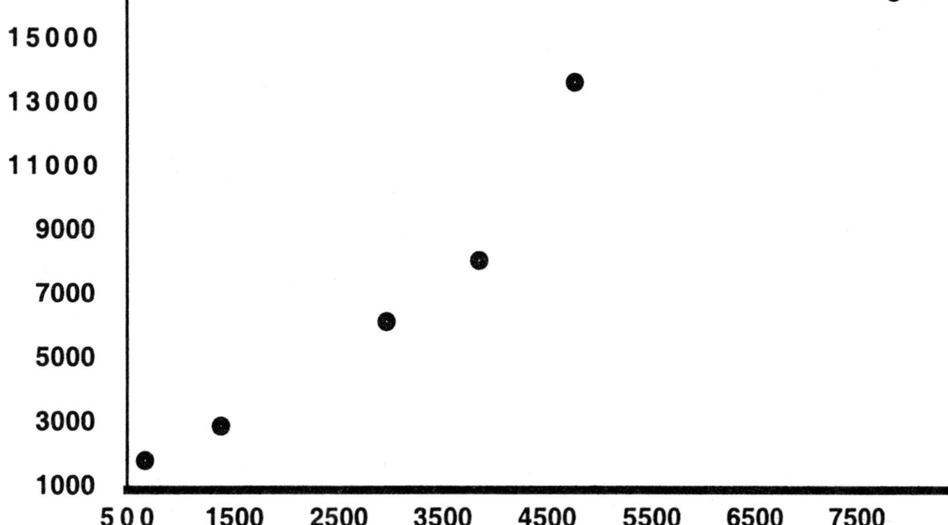

b. Is there a pattern displayed by the points on the scattergram? Explain your answer. ***(12)***

The explanation might contain information such as: It appears that the number of male coaches and the number of female coaches are in a linear pattern except for Pennsylvania. Or the number of male coaches is about twice the number of female coaches except in Pennsylvania where the number is about three times.

2. MADD (Mothers Against Drunk Driving) is a national organization that provides information and educational programs to assist local groups and governments to provide programs to make the public aware of the problems with drunk drivers. Melissa Hargrove, president of the local chapter of MADD, researched the fatalities from automobile accidents from 1966 to 1990. She presented the following data to her chapter.

TRAFFIC FATALITIES

YEAR	FATALITY RATES PER 100,000 DRIVERS
1966	50
1970	47
1974	36
1978	36
1982	29
1986	29
1990	27

The scattergram for this data is displayed below:

TRAFFIC FATALITIES / 100,000 DRIVERS

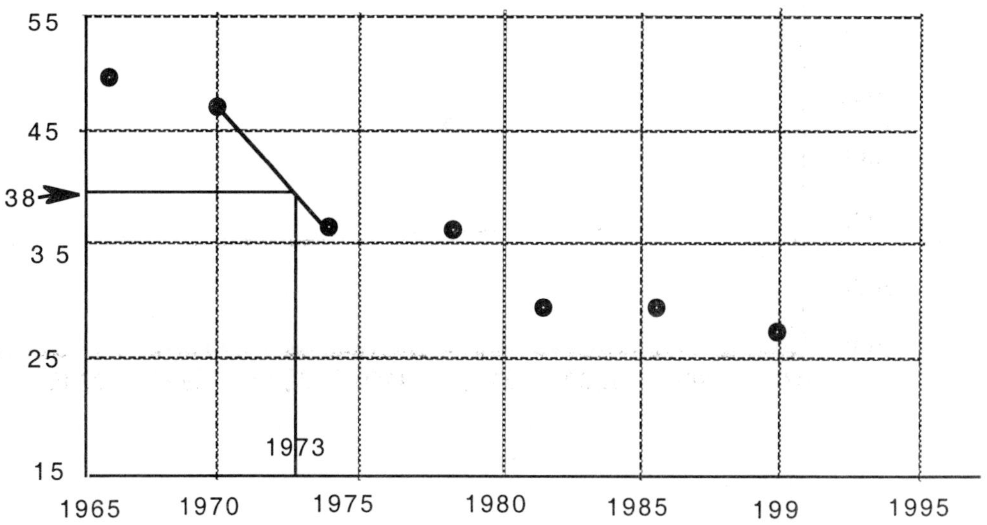

a. Lucy Winfield wondered what the fatality rate was in 1973. In order to find this from the graph, linear interpolation was used. Use a ruler to demonstrate how you can find an estimate for the number of fatalities in 1973 using linear interpolation. What is your estimate? ___*about 38*___ *See Graph.* ***(15)***

b. When using linear interpolation, what does your answer represent? ***(6)***

Linear interpolation allows you to find an estimate, in this case, of a y value between two given x values. Using linear interpolation in this problem, we are able to estimate the number of deaths for 1973 when we know the deaths for 1970 and 1974.

c. Find the equation of the line that contains the line segment connecting (70, 47) to (74, 36). Show your work. ***(12)***

$$y = \frac{-11}{4}x + 239.5$$

d. Use the equation from part c. to find an estimate for the number of fatalities in 1973 by substituting 73 into the equation. Show your work. ***(12)***

$$y = \frac{-11}{4} * 73 + 239.5$$

$$y = 38.75 \text{ deaths}$$

e. How did the your estimates from parts a. and d. compare? Be as specific as you scan. ***(6)***

Student answers will vary.

f. Write at least two reasons why you think fatalities decreased from 1966 to 1990. ***(12)***

A rich answer could contain:

1. *There was a decrease in the speed limit to 55 mph.*
2. *There has been an increase of safety features in cars.*
3. *There has been a campaign against drunk drivers.*
4. *There has been an increase in education about safe driving.*

MATH *Connections* **I**
GRAPHICAL ESTIMATION QUIZ
SECTIONS 4.3 - 4.5 (A)

Name ________________________________ Date ________________

Melissa Hargrove, president of the local chapter of MADD, finds you are able to use a calculator to analyze data that is displayed on a scattergram. Since she needs more information for a media conference for the local newspaper and television studios, she asks you to work with her in preparing materials. The data she wishes to analyze is in the table and scattergram below:

TRAFFIC FATALITIES

YEAR	FATALITY RATES PER 100,000 DRIVERS
1966	50
1970	47
1974	36
1978	36
1982	29
1986	29
1990	27

TRAFFIC FATALITIES / 100,000 DRIVERS

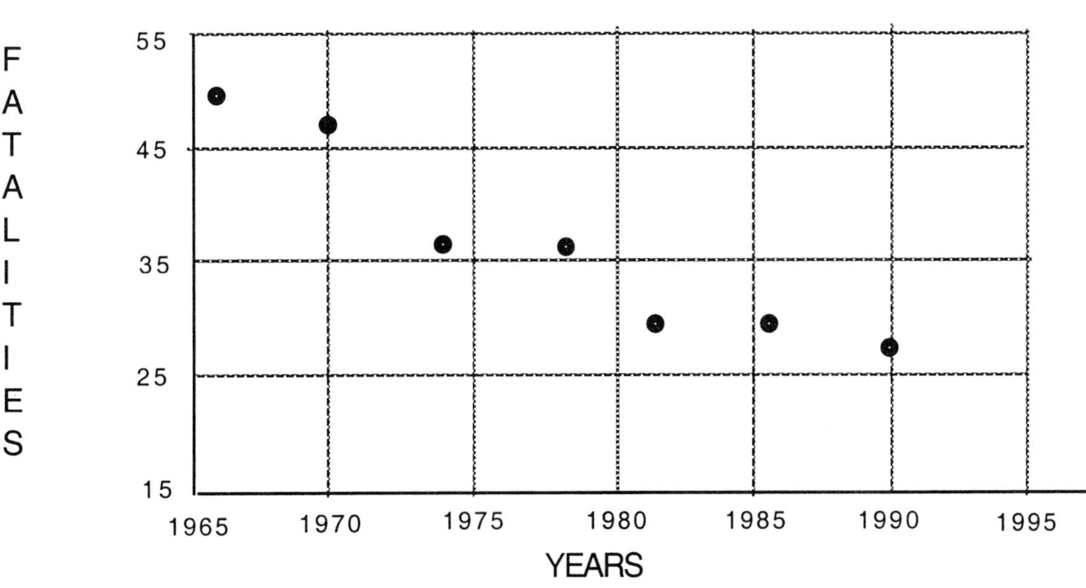

1. a. Draw a line on the graph on page 1, that you feel is the best representation of the data. Explain how you decided to draw your line.

____

____

____

b. You enter the data by choosing the STAT menu, selecting EDIT and keying in the data in L_1 and L_2. (Enter the dates with the last two digits; 1966 as 66.)

c. You then choose the STAT menu and select the CALC option followed by menu choice 5 (LINREG(AX+B) L1,L2) (for the TI-83 choice 4) to determine the slope and Y-intercept of the least-squares line for the data:

Slope: ______ Y-intercept _____

Equation of least-squares line: ______________________

d. Explain how the least-squares line is determined.

____

____

____

____

e. Melissa would like to know the approximate number of fatalities in 1973, the year after the 55 MPH speed limit became law. You use your calculator to evaluate the least-squares equation when x is 73. Your answer is _______.
Plot this point on the graph below.

Use the graph for the answers to 1e, 1f, 1g.

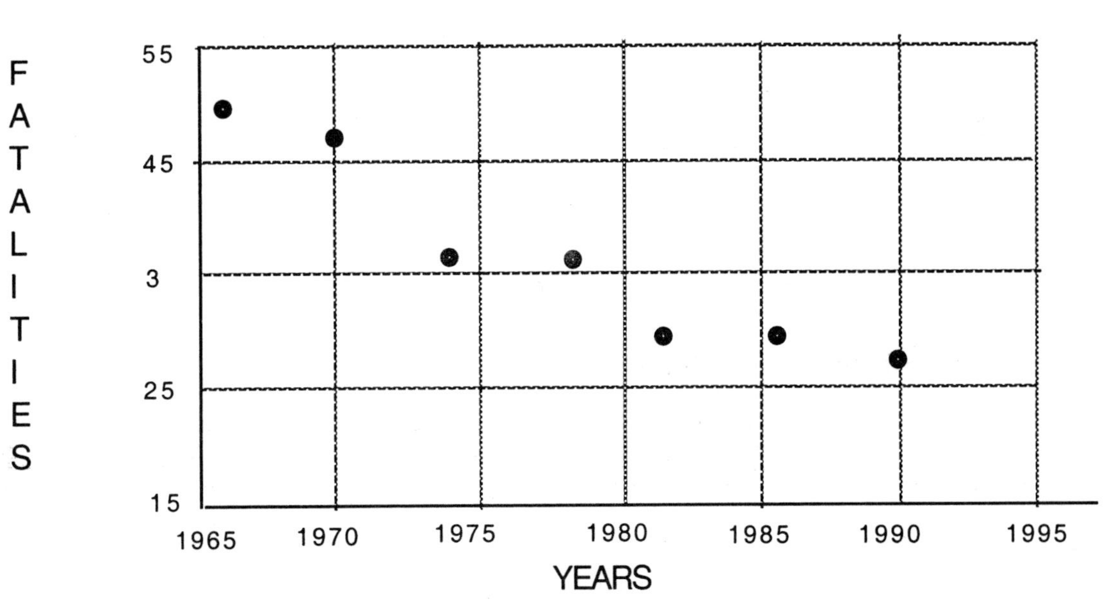

f. Many wish to know what the forecast about fatalities will be for 1995. You find it by substituting 95 in your equation. Your answer is ________
Plot this point on the graph above.

g. You connect the two points you just found to display the least-squares line. Using this line and not the calculator, make estimates for 1979 and 1989.
Your estimates are: 1979 _________ 1989 _________

h. Melissa asks you to speak about the data and scattergram at the media conference. What would you say?

____

____

____

____

____

____

____

____

2. Explain the difference between interpolation and extrapolation.

____

____

____

____

____

3. List two limitations of extrapolation.

____

____

____

____

SOLUTION KEY AND SCORING GUIDE

MATH *Connections* I
GRAPHICAL ESTIMATION QUIZ
SECTIONS 4.3 - 4.5 (A)

Melissa Hargrove, president of the local chapter of MADD, finds you are able to use a calculator to analyze data that is displayed on a scattergram. Since she needs more information for a media conference for the local newspaper and television studios, she asks you to work with her in preparing materials. The data she wishes to analyze is in the table and scattergram below:

TRAFFIC FATALITIES

YEAR	FATALITY RATES PER 100 000 DRIVERS
1966	50
1970	47
1974	36
1978	36
1982	29
1986	29
1990	27

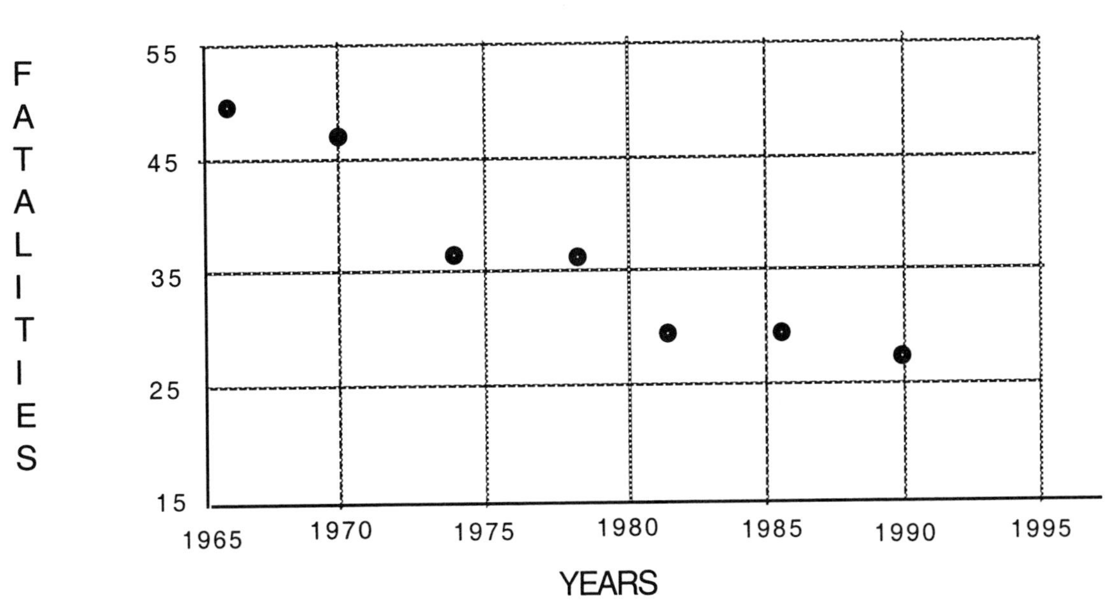

1. a. Draw a line on the graph on page 1. that you feel is the best representation of the data. Explain how you decided to draw your line. ***(12)***

The line will be the student's best estimate. Full credit would be given for a line that falls into the "middle" of the points. Explanations will vary. Students should mention the distances of points to line as part of the explanation.

b. You enter the data by choosing the STAT menu, selecting EDIT and typing the data in L_1 and L_2. (Enter the dates with the last two digits; 1966 as 66.)

c. You then choose the STAT menu and choose the CALC option followed by menu choice 5 (LINREG(AX+B) L1,L2) (for the TI-83 choose 4) to determine the slope and Y-intercept of the least-squares line for the data:

Slope: <u>*- 1*</u> ***(8)*** Y-intercept <u>*114*</u> ***(8)***

Equation of least-squares line: <u>*y = -1 x + 114*</u> ***(8)***

d. Explain how the least-squares line is determined. ***(6)***
Student should explain how distances from data points to lines are found. The line, for which the sum of the squares of the distances is least, is determined to be the least-squares line.

e. Melissa would like to know the approximate number of fatalities in 1973, the year after the 55 MPH speed limit became law. You use your calculator to evaluate the least-squares equation when x is 73. Your answer is _41_. ***(10)***
Plot this point on the graph below.

Use the graph for the answers to 1e, 1f, 1g.

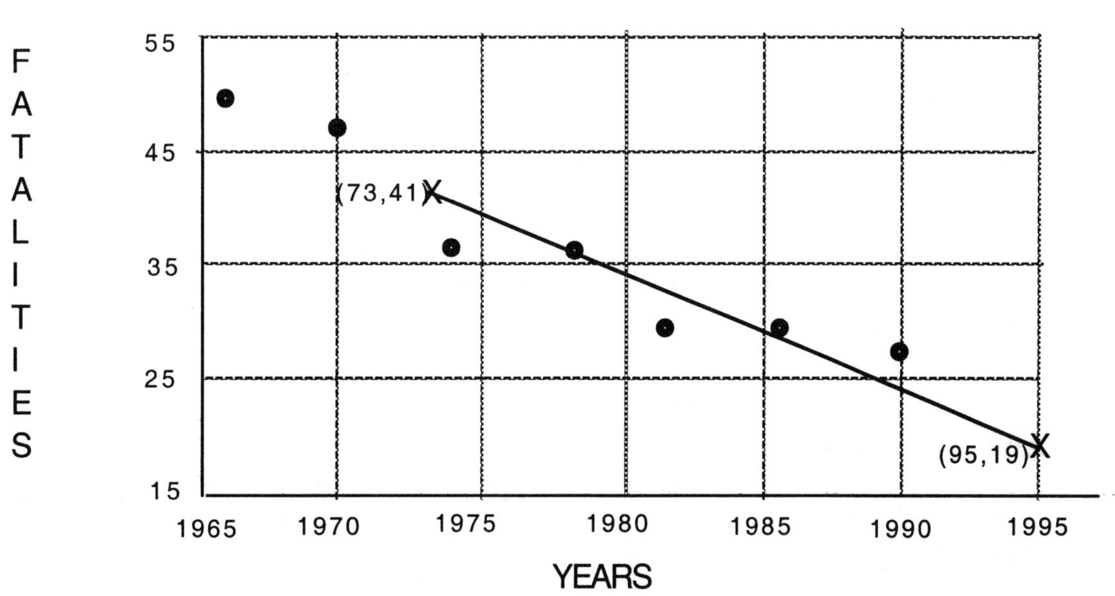

f. Many wish to know what the forecast about fatalities will be for 1995. You find it by substituting 95 in your equation. Your answer is _19_ ***(8)***
Plot this point on the graph above.

g. You connect the two points you just found to display the least-squares line. Using this line and not the calculator, make estimates for 1979 and 1989.
Your estimates are: 1979 _35_ ***(8)*** 1989 _25_ ***(8)***

h. Melissa asks you to speak about the data and scattergram at the media conference. What would you say? ***(12)***

A rich answer could include the following:

- *The data and scattergram show a pattern of decreasing fatalities. It is nearly a 50% decrease.*
- *Using a line that comes "close" to the data, it can be forecast that fatalities could continue to decrease. Forecasts for many years from now cannot be accurate until recent trends are clear.*
- *In recent years the line that comes "close" to the data has been below the actual data.*

2. Explain the difference between interpolation and extrapolation. ***(6)***

Interpolation is the process of finding data between known data while extrapolation is finding data that is outside the given data. When one is predicting, then extrapolation would be used.

3. List two limitations of extrapolation? ***(6)***

   1. *If data is always decreasing one would be led to predict information that might be physically impossible to achieve.*
   2. *Unanticipated changes may occur that make the data that is being used to predict things that are occurrences that are unrelated to the present situation.*

MATH *Connections* **I**
GRAPHICAL ESTIMATION TEST (A)
CHAPTER 4

Name ______________________________ Date ______________

1. You are a summer intern working with Harry Kenworthy from Rogers Corporation who is doing some research for the purchasing department. He researched the following data involving ice cream sales during two-week intervals during the year. The data is listed below.

ICE CREAM CONSUMPTION

CALENDAR WEEKS	AVG. BI-WEEKLY TEMP ° F	NO. OF GALLONS SOLD
1-2	28	170
3-4	28	160
5-6	29	175
7-8	34	225
9-10	36	230
11-12	42	230
13-14	44	250
15-16	52	280
17-18	64	325
19-20	64	335
21-22	70	405
23-24	71	420
25-26	77	450
27-28	80	450
29-30	81	490
31-32	82	500
33-34	84	495
35-36	80	465
37-38	72	450
39-40	65	400
41-42	60	325
43-44	56	250
45-46	46	250
47-48	43	225
49-50	39	190
51-52	40	180

Harry made the scattergram below to see if there were any trends or patterns:

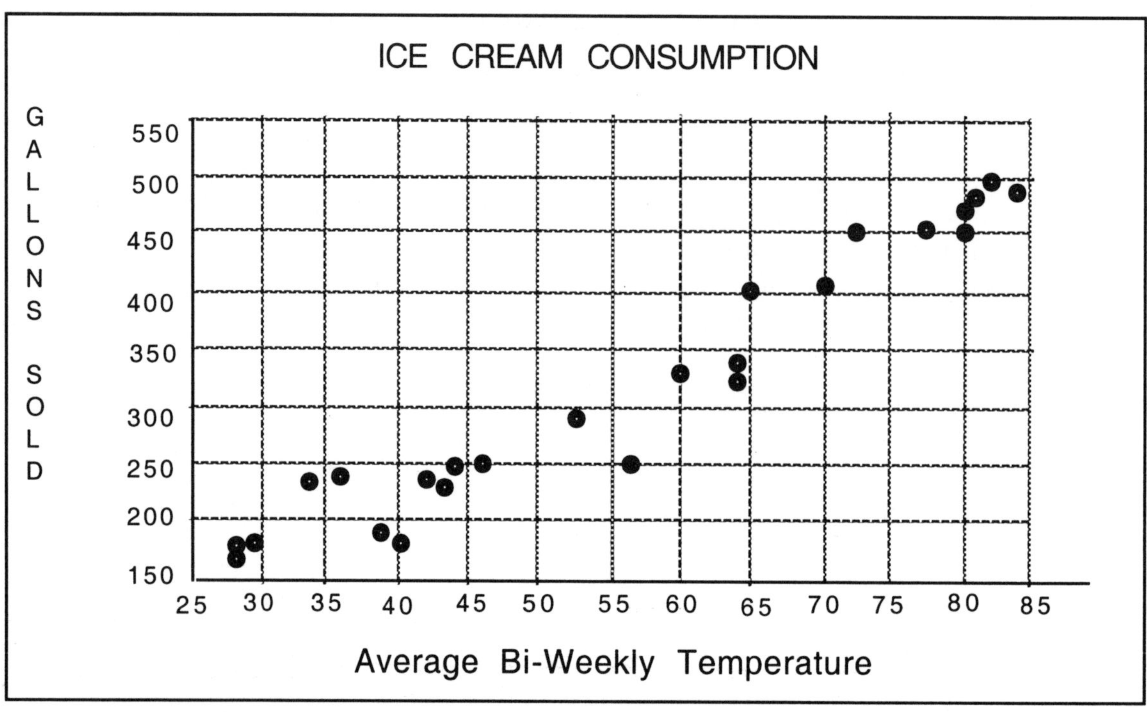

a. After completing the scattergram Harry noticed he left out the data for the 23-24 weeks interval. Place it on the graph for him with an X.

b. Harry asks you what trend or pattern you see on the scattergram. Your response is:

____

____

____

____

____

c. Harry receives a call from a client in Denver who tells him that the bi-weekly temperature is expected to average 59°. Harry asks you to use linear interpolation to estimate the number of gallons that would be sold at that temperature. You discover that the batteries on your calculator are dead so you can only use a ruler to find the estimate. Show on the graph how you completed your work. Estimate: ________________

d. A client from Houston calls and tells Harry that a prolonged heat wave has been forecast and she needs an estimate of how many gallons of ice cream she should purchase if the average bi-weekly temperature is 90°. Harry again asks you to do the work. Show it on the graph. Estimate: ____________________

e. That afternoon you purchase new batteries. Now you can use your calculator functions to find the estimates you were forced to do by hand earlier in the day. First you select the STAT menu, choose EDIT and enter the data in L_1 and L_2. Then from the STAT menu, select the CALC option and choose (LINEREG(AX+B) L_1,L_2) to determine the slope and y-intercept of the least-squares line for the data. Slope _______ y-intercept _______

Equation of least-squares line: y = ______________________________

f. Use the equation in e. and your calculator to estimate the number of gallons sold when temperatures are 59° and 90°.

59° __________ 90° __________

Discuss how well your estimates in parts c. and d. compare to your estimates using the least-squares equation.

____

____

____

g. If there is a severe cold wave and the average bi-weekly temperature is 0°F, explain what the estimate from the least-squares equation means.

____

____

____

2. At Park School the principal decided to make a study of the average grade on the MATH *Connections* Final Exam and the number of students in each of the ten classes. This is the data she collected:

Park School
MATH *Connections* Final Exam

Teacher	# of Students	Mean of the Grades
Mr. Compy	23	72
Miss Brontus	28	79
Mrs. Brachio	41	67
Mr. Apato	35	75
Ms Stego	20	80
Ms Trice	25	76
Mr. Micro	31	73
Dr. Ichthy	37	72
Miss Hypsil	19	79
Mr. Quetzo	36	71

a. Ms. Dipaldo, the principal, makes a scattergram and draws in the line she feels comes "close" to the data. That is below.

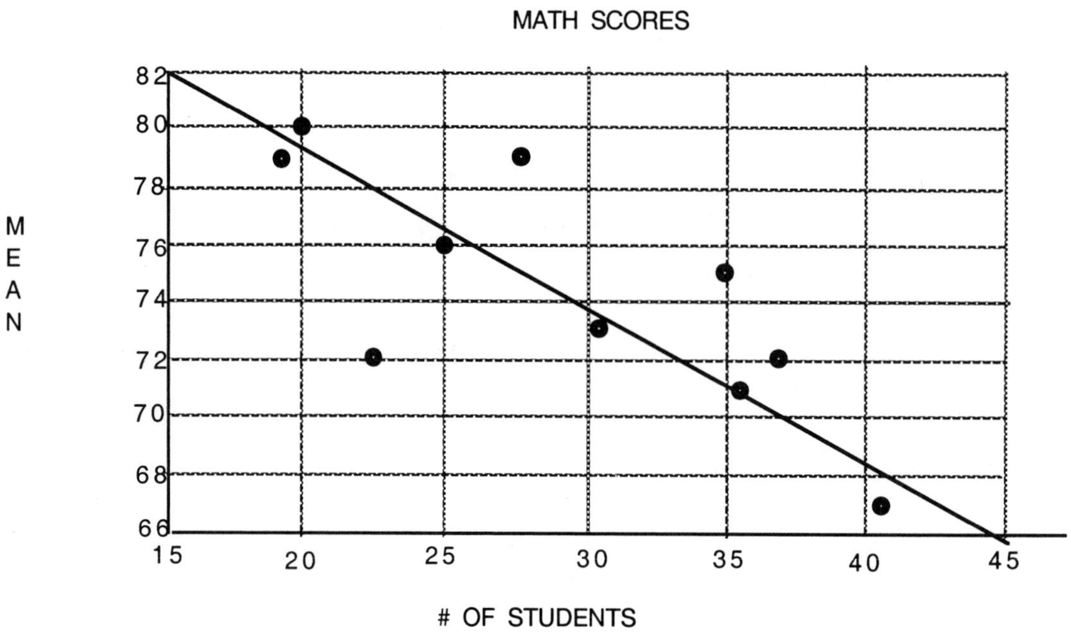

She asks you to comment on the trend or pattern that you see in the scattergram. Your reply is:

____

____

____

____

____

b. Ms. Dipaldo wants to check if the line she drew is "close" to the data. She needs to find the vertical distances d_1 through d_{10} which represent the difference in means from each data point to the line she drew. She starts the following table and asks you to complete it. **She is estimating the distances to the nearest .5 unit.**

	d	d^2		d	d^2
d_1	1	1	d_6		
d_2	1	1	d_7		
d_3	6	36	d_8		
d_4	.5		d_9		
d_5			d_{10}		

To complete the analysis, Ms. Dipaldo has to find the following three tests that are used to determine if the line is a close fit in relation to another line. Find the three values from the information in the table.

i. $d_1 + d_2 + d_3 + d_4 + d_5 + d_6 + d_7 + d_8 + d_9 + d_{10} =$ ______

ii. $d_1{}^2 + d_2{}^2 + d_3{}^2 + d_4{}^2 + d_5{}^2 + d_6{}^2 + d_7{}^2 + d_8{}^2 + d_9{}^2 + d_{10}{}^2 =$ ______________

iii. Largest d = _____

iv. Discuss the three values you just determined. Explain what they are and how they are used.

____

____

____

____

____

v. Which of the three tests which do you feel is the most accurate. Explain your answer.

EXTRA CREDIT:

c. A challenge would be to determine if you could find a line that fits the data better than the one Ms. Dipaldo drew. Use the graph below to find a line of best fit that is better than the one Ms. DiPaldo drew.

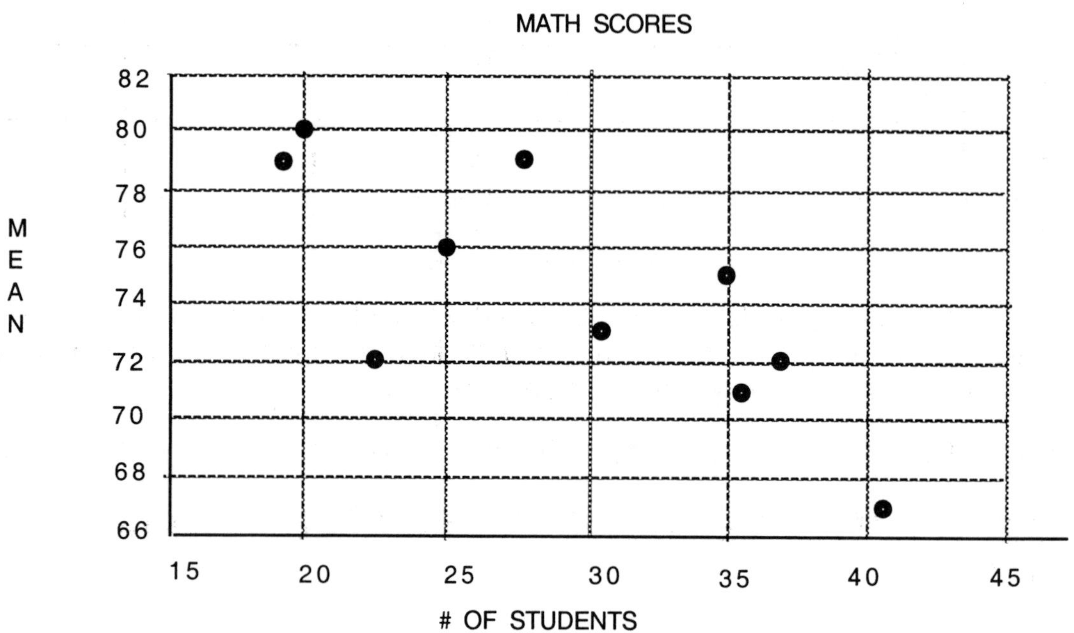

SOLUTION KEY AND SCORING GUIDE

MATH *Connections* **I**
GRAPHICAL ESTIMATION TEST (A)
CHAPTER 4

1. You are a summer intern working with Harry Kenworthy from Rogers Corporation who is doing some research for the purchasing department. He researched the following data involving ice cream sales during two-week intervals during the year. The data is listed below:

ICE CREAM CONSUMPTION

CALENDAR WEEKS	AVG. BI-WEEKLY TEMP ° F	NO. OF GALLONS SOLD
1-2	28	170
3-4	28	160
5-6	29	175
7-8	34	225
9-10	36	230
11-12	42	230
13-14	44	250
15-16	52	280
17-18	64	325
19-20	64	335
21-22	70	405
23-24	71	420
25-26	77	450
27-28	80	450
29-30	81	490
31-32	82	500
33-34	84	495
35-36	80	465
37-38	72	450
39-40	65	400
41-42	60	325
43-44	56	250
45-46	46	250
47-48	43	225
49-50	39	190
51-52	40	180

Harry made the scattergram below to see if there were any trends or patterns:

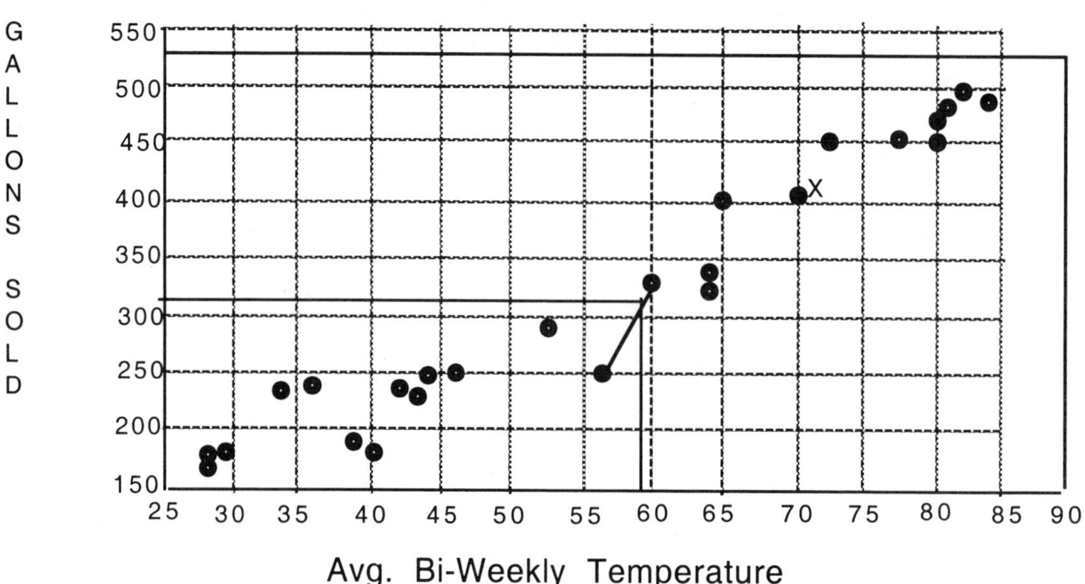

a. After completing the scattergram, Harry noticed he left out the data for the 23-24 weeks interval. Place it on the graph for him with an X. ***(5)***

b. Harry asks you what trend or pattern you see on the scattergram. Your response is: ***(6)***

> *As the daily temperature increases the number of gallons of ice cream sold increases.*

c. Harry receives a call from a client in Denver who tells him that the bi-weekly temperature is expected to average 59°. Harry asks you to use linear interpolation to estimate the number of gallons that would be sold at that temperature. You discover that the batteries on your calculator are dead, so you can only use a ruler to find the estimate. Show on the graph how you completed your work. Estimate: *310 gallons* ***(6)***

d. A client from Houston calls and tells Harry that a prolonged heat wave has been forecast, and she needs an estimate of how many gallons of ice cream she should purchase if the average bi-weekly temperature is 90°. Harry again asks you to do the work. Show it on the graph. Estimate: *530 gallons* ***(6)***

e. That afternoon you purchase new batteries, so now you can use your calculator functions to find the estimates you were forced to do by hand earlier in the day. First you select the STAT menu, choose EDIT and enter the data in L_1 and L2. Then from the STAT menu, select the CALC option and choose (LINEREG(AX+B) L_1,L_2) to determine the slope and y-intercept of the least-squares line for the data. Slope _6_ ***(5)*** y-intercept _- 20_ ***(5)***

Equation of least squares line: y = _6x –20_ ***(5)***

f. Use this equation and your calculator to estimate the number of gallons sold if the temperatures are 59° and 90°.

59° _334_ ***(5)*** 90° _520_ ***(5)***

Discuss how well your estimates in parts c. and d. compare to your estimates using the least-squares equation. ***(6)***

Students should note that these were estimates and give some rationale for the differences.

g. If there is a severe cold wave and the average bi-weekly temperature is 0°F, explain what the estimate from the least-squares equation means. ***(6)***

It means you would sell less than zero gallons of ice cream, which does not make any sense. Students should discuss the limitations of the algebraic model.

2. At Park School, the principal decided to make a study of the average grade on the MATH *Connections* Final Exam and the number of students in each of the ten classes. This is the data she collected:

Park School
MATH *Connections* Final Exam

Teacher	# of Students	Mean of the Grades
Mr. Compy	23	72
Miss Brontus	28	79
Mrs. Brachio	41	67
Mr. Apato	35	75
Ms Stego	20	80
Ms Trice	25	76
Mr. Micro	31	73
Dr. Ichthy	37	72
Miss Hypsil	19	79
Mr. Quetzo	36	71

a. Ms. Dipaldo, the principal, makes a scattergram and draws in the line she feels comes "close" to the data. That is shown below.

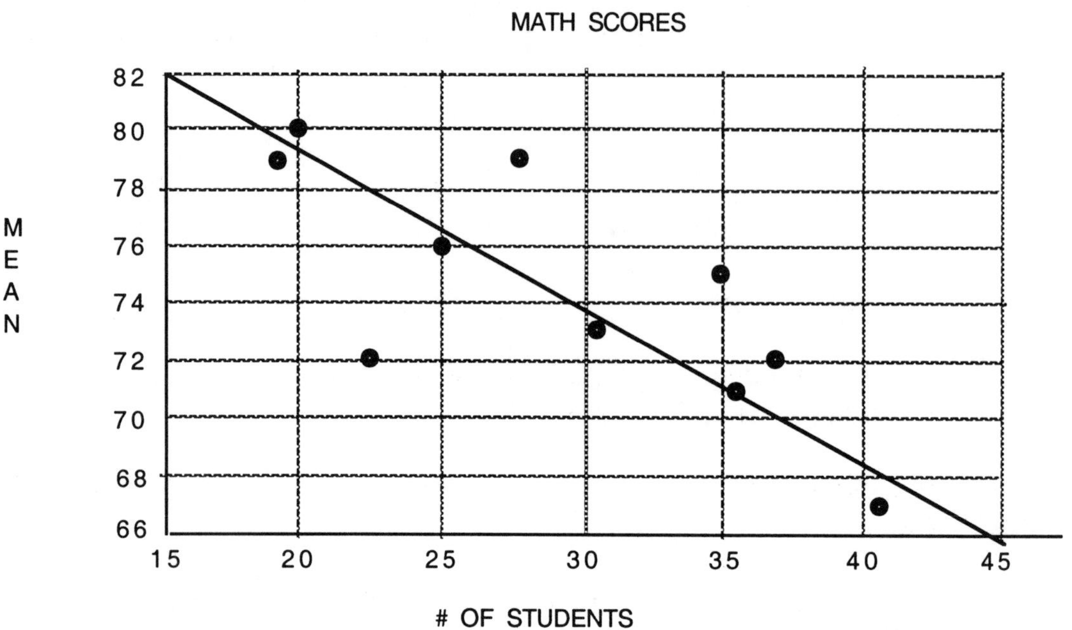

She asks you to comment on the trend or pattern that you see in the scattergram. Your reply is: ***(6)***

As the number of students in a class increases the mean for the test results decreases.

b. Ms. Dipaldo wants to check if the line she drew is "close" to the data. She needs to find the vertical distances d_1 through d_{10} which represent the difference in means from each data point to the line she drew. She starts the table below and asks you to complete it. **She is estimating the distances to the nearest .5 unit.** ***(7)***

	d	d2		d	d2
d_1	1	1	d_6	.5	.25
d_2	1	1	d_7	4	16
d_3	6	36	d_8	0	0
d_4	.5	.25	d_9	2	4
d_5	4	16	d_{10}	1	1

To complete the analysis, Ms. Dipaldo has to find the following three tests that are used to determine if the line is a close fit in relation to another line. Find the three values from the information in the table.

i. $d_1 + d_2 + d_3 + d_4 + d_5 + d_6 + d_7 + d_8 + d_9 + d_{10} =$ *20* ***(5)***

ii. $d_1{}^2 + d_2{}^2 + d_3{}^2 + d_4{}^2 + d_5{}^2 + d_6{}^2 + d_7{}^2 + d_8{}^2 + d_9{}^2 + d_{10}{}^2 =$ *75.5* ***(5)***

iii. Largest d = *6* **(5)**

iv. Discuss the three values you just determined. Explain what they are and how they are used. ***(6)***

The discussion should clearly identify (i) as the sum of the differences, (ii) as the sum of the squares of the differences and (iii) as the largest of the 10 differences. The student should also discuss the fact that the tests are used to determine if on line of best fit is a "better " fit than another line. Some historical background would be a plus.

v. Which of the three tests which do you feel is the most accurate. Explain your answer. ***(6)***

This is clearly an open-ended question that asks the student to take a position and defend it. There is no one answer to this question.

EXTRA CREDIT:

c. A challenge would be to determine if you could find a line that fits the data better than the one Ms. Dipaldo drew. Use the graph below to find a line of best fit that is better than the one Ms. DiPaldo drew.

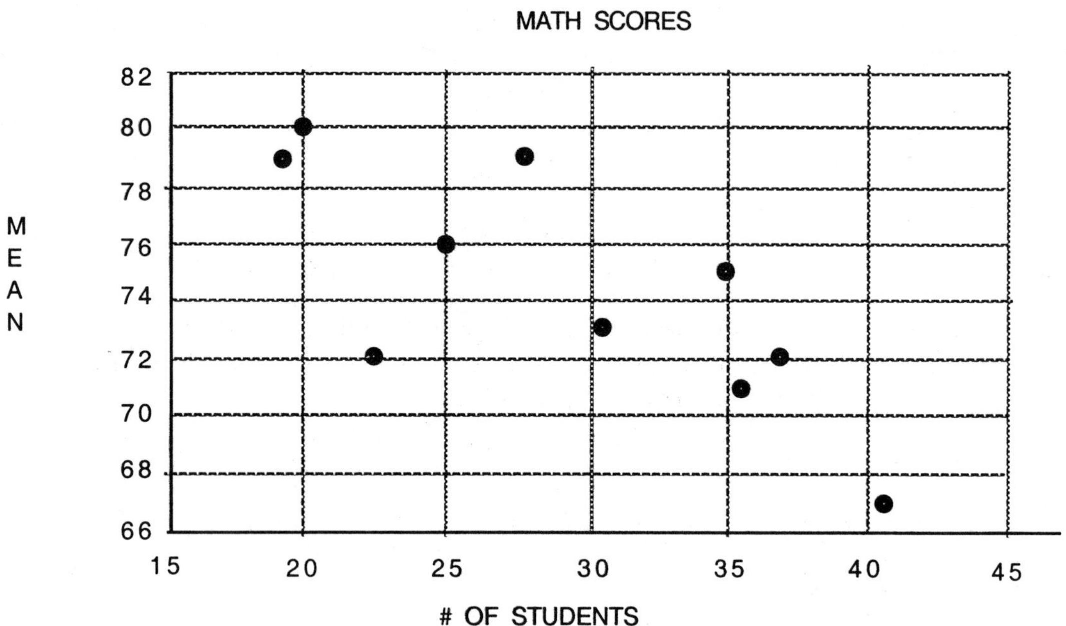

c. *Student answers will vary.*